BEYOND ENVIRONMENTAL LAW

This book offers a vision for the third generation of environmental law designed to enhance its ability to protect our environment. The book presents two core proposals, an Environmental Legacy Act, to preserve a defined environmental legacy for future generations, and an Environmental Competition Statute, to spark movement to new clean technologies. The first proposal would require, for the first time, that the federal government define an environmental legacy that it must preserve for future generations. The second would establish a market competition to maximize environmental protection. The balance of the book provides complementary proposals and analysis. The first generation of environmental law sought broad protection of health and the environment in a fairly fragmented way. The second sought to enhance environmental law's efficiency through cost-benefit analysis and market mechanisms. These proposals seek to create a broader, more creative approach to solving environmental problems.

Alyson C. Flournoy is Professor and Alumni Research Scholar at the University of Florida's Levin College of Law, where she also serves as Director of the Environmental and Land Use Law Program. Her writing focuses on decision-making processes under environmental and natural resource laws, environmental ethics, and the intersection of science and law and has appeared in leading journals including *Columbia Journal of Environmental Law*, *Harvard Environmental Law Review*, *Indiana Law Journal*, and *Texas Law Review*. Flournoy is also a member-scholar of the Center for Progressive Reform.

David M. Driesen is the thirteenth professor in Syracuse University's history to hold the title University Professor, the highest honor available to a faculty member at that institution. His research focuses on the law and economics of environmental law. His book *The Economic Dynamics of Environmental Law* (2003) won a Lynton Keith Caldwell Award for the best book on environmental and technology policy. He has published a textbook, *Environmental Law: A Conceptual and Pragmatic Approach* (2007) (with Robert Adler), and his articles have appeared in leading journals, including *Cornell Law Review*, *Ecology Law Quarterly*, *Harvard Environmental Law Review*, and *Virginia Journal of International Law*. Driesen is also a member-scholar of the Center for Progressive Reform.

Beyond Environmental Law

POLICY PROPOSALS FOR A BETTER ENVIRONMENTAL FUTURE

Edited by

Alyson C. Flournoy

Levin College of Law, University of Florida

David M. Driesen

College of Law, Syracuse University

CAMBRIDGE
UNIVERSITY PRESS

CAMBRIDGE UNIVERSITY PRESS
Cambridge, New York, Melbourne, Madrid, Cape Town, Singapore,
São Paulo, Delhi, Dubai, Tokyo

Cambridge University Press
32 Avenue of the Americas, New York, NY 10013-2473, USA

www.cambridge.org
Information on this title: www.cambridge.org/9780521744324

First published 2010

Printed in the United States of America

A catalog record for this publication is available from the British Library.

Library of Congress Cataloging in Publication data

Beyond environmental law : policy proposals for a better environmental future /
edited by Alyson C. Flournoy, David M. Driesen.
p. cm.
Includes bibliographical references and index.
ISBN 978-0-521-76771-2 (hardback) – ISBN 978-0-521-74432-4 (pbk.)
1. Environmental law – United States. 2. Pollution – Law and legislation – United States.
3. Environmental protection – United States – Standards. 4. Green technology –
Government policy – United States. I. Flournoy, Alyson, C., 1958– II. Driesen, David M.
KF3775.B435 2010
344.7304′6 – dc22 2009034835

ISBN 978-0-521-76771-2 Hardback
ISBN 978-0-521-74432-4 Paperback

For John

– A.C.F.

For Kim, Mai, and Mirrah

– D.M.D.

Contents

List of Figure and Tables

FIGURE

TABLES

About the Contributors

David E. Adelman is a Professor at the University of Texas School of Law. Before joining the Texas faculty in 2009, Adelman was an associate professor of law at the James E. Rogers College of Law. Before becoming a law professor, he was a senior attorney for the Natural Resources Defense Council in Washington, D.C., as well as an associate at the Washington office of Covington & Burling. Adelman has authored such articles as "Scientific Activism and Restraint: The Interplay of Statistics, Judgment, and Procedure in Environmental Law" (*Notre Dame Law Review* 2004); "The False Promise of the Genomics Revolution for Environmental Law" (*Harvard Environmental Law Review* 2005); and "A Fallacy of the Commons in Biotech Patent Policy" (*Berkeley Technology Law Journal* 2005). Adelman received his J.D. from Stanford Law School, where he participated in the *Stanford Environmental Law Journal.* Adelman also received his Ph.D. in chemical physics from Stanford University and his B.A. from Reed College.

Mary Jane Angelo is an Associate Professor at the University of Florida Levin College of Law. Her teaching and scholarship are focused in the areas of environmental law, water law, administrative law, biotechnology law, dispute resolution, pesticides law, and law and science. Before joining the University of Florida, Angelo served as the senior assistant general counsel at the St. Johns River Water Management District and as an attorney for the U.S. Environmental Protection Agency, in Washington, D.C. Angelo has published several articles on environmental issues, including "The Killing Fields: Reducing the Casualties in the Battle between U.S. Endangered Species and Pesticide Law" (*Harvard Environmental Law Review*

2008); "Regulating Evolution for Sale: An Evolutionary Biology Model for Regulating the Risks Posed by Genetically Modified Organisms" (*Wake Forest Law Review* 2007); and "Embracing Uncertainty, Complexity and Change to Protect Ecological Integrity: An Eco-Pragmatic Reinvention of a First Generation Environmental Law" (*Ecology Law Quarterly* 2006). Angelo received her J.D. and M.S. from the University of Florida and her B.A. from Rutgers University.

Thomas T. Ankersen is a Legal Skills Professor and Director of the Conservation Clinic at the University of Florida Levin College of Law. He is also the Director of the University of Florida/University of Costa Rica Joint Program in Environmental and Land Use Law in Costa Rica. Before joining the University of Florida, Ankersen served as a staff attorney for the Sierra Club National Defense Fund and as a senior litigation associate at Peeples, Earl & Blank. Ankersen has authored many articles and book chapters in the area of environmental law, including "Shared Knowledge; Shared Jurisprudence: Learning to Speak Environmental Law Creole" (*Tulane Law Review* 2003) and "Inside the Polygon: Emerging Community Tenure Systems" and *Forest Resource Extraction in Working Forests in the Tropics: Conservation Through Sustainable Management* (Columbia University Press 2004). Ankersen received his J.D. from the University of Florida and his M.A. and B.A. from the University of South Florida.

Mark T. Brown is a Professor in environmental engineering sciences and the Director of the Center for Environmental Policy at the University of Florida. Before becoming a professor, Brown was a research scientist and associate program director with the University of Florida's Howard T. Odum Center for Wetlands. Brown's current research includes projects to develop ecological indicators of wetland ecosystem health; the development of indices of success for restored wetlands; the restoration of drastically altered landscapes; and quantitative evaluation of natural capital and environmental services. In addition to his research studies, Brown has coauthored many articles, including "Predicting National Sustainability: The Convergence of Energetic, Economic and Environmental Realities" (*Ecological Modeling* 2009); "Emergy and Ecosystem Complexity" (*Communications in Nonlinear Science and Numerical Simulation* 2009); and "Species Diversity in the Florida Everglades, USA: A Systems Approach to Calculating Biodiversity" (*Aquatic Sciences* 2007). Brown received his Ph.D., M.A., and B.A. from the University of Florida.

David M. Driesen is University Professor at Syracuse University. He has taught environmental law (domestic and international) and constitutional law. Before joining the Syracuse College of Law, Driesen worked in Washington, D.C., as a Senior Project Attorney for the Natural Resources Defense Council, Air and Energy Program. His publications include *The Economic Dynamics of Environmental Law* (MIT Press 2003), *Environmental Law: A Conceptual and Pragmatic Approach* (Aspen 2007) (with Robert Adler), "Is Cost-Benefit Analysis Neutral?" (*Colorado Law Review* 2006), "Standing for Nothing: The Paradox of Demanding Concrete Context for Formalist Adjudication" (*Cornell Law Review* 2004), and "What Is Free Trade? The Real Issue Lurking behind the Trade and Environment Debate" (*Virginia Journal of International Law* 2001). Driesen received his J.D. from the Yale Law School, his M.Mus. from the Yale School of Music, and his B.Mus. from the Oberlin Conservatory.

Alyson C. Flournoy is a Professor of Law and Alumni Research Scholar and the Director of the Environmental and Land Use Law Program at the University of Florida Levin College of Law. Flournoy's teaching and scholarship are focused in the areas of environmental law, property, administrative law, and advanced environmental law and litigation. Before joining the University of Florida, Flournoy served as an associate attorney for Covington & Burling in the Washington, D.C., office. Flournoy has authored several book chapters and articles in the area of environmental law, including "Protecting a Natural Resource Legacy While Promoting Resilience: Can It Be Done?" (*Nebraska Law Review* 2009); "Harnessing the Power of Information to Protect Our Public Natural Resource Legacy" (*Texas Law Review* 2008) (with Heather Halter and Christina Storz); and "Supply, Demand, and Consequences: The Impact of Information Flow on Individual Permitting Decisions under Section 404 of the Clean Water Act" (*Indiana Law Journal* 2008). Flournoy received her J.D. from Harvard Law School and her A.B. from Princeton University.

Christine A. Klein is a Chesterfield Smith Professor of Law in the Environmental and Land Use Law Program at the University of Florida Levin College of Law. Klein's teaching and scholarship are focused in the areas of natural resources law, water law, and property. Before joining the University of Florida, Klein was chair of the Environmental Law Concentration Program and a professor at Michigan State University College of Law. She also served as an assistant attorney general of the Natural Resources Section for the State of Colorado. Professor Klein has published articles in her area of practice, including "The Environmental Deficit: Applying Lessons from the Economic Recession" (*Arizona Law Review* 2009), "Modernizing Water Law: The Example of Florida" (*Florida Law Review* 2009) (with Mary Jane Angelo and Richard Hamann), and "Cultural Norms as a Source of Law: The Example of Bottled Water" (*Cardozo Law Review* 2008) (with Ling-Yee Huang). She has also coauthored the book *Natural Resources Law: A Place-Based Book of Problems and Cases* (Aspen Publishers 2nd ed. 2009). Klein received her L.L.M. from Columbia University School of Law, her J.D. from the University of Colorado, and her B.A. from Middlebury College.

Brian Mayer is an Assistant Professor in the Department of Sociology and the Department of Epidemiology and Biostatistics at the University of Florida. Mayer's teaching and scholarship are focused in the areas of environmental sociology, social movements, and medical sociology. He is the author of *Blue-Green Coalitions: Fighting for Safer Workplaces and Healthy Communities* (Cornell University Press 2008). Mayer has also authored several articles and book chapters in the field of environmental sociology, including "Cross-Movement Coalition Formation: Bridging the Labor-Environment Divide" (2009), "School Custodians and Green Cleaners: New Approaches to Labor-Environmental Coalitions" (2007), and "Clearing the Air and Breathing Freely: The Health Politics of Air Pollution and Asthma" (2004). He also sits on the board of several environmental organizations, including the Alliance for a Healthy Tomorrow and the Toward Tomorrow Project, and is a senior Fellow with the Environmental Leadership Program. Mayer received his M.A. and Ph.D. from Brown University and his B.A. from the University of California, Santa Cruz.

Christine Overdevest is an Assistant Professor in the Department of Sociology and an affiliate faculty of the School of Natural Resources at the University of Florida. In 2008–2009 she was a visiting researcher at the Institute for International Business at the Stockholm School of Economics. Overdevest's teaching and scholarship are focused in the areas of environmental governance and economic sociology. Professor Overdevest has published articles in her area of expertise, including a contribution to a special issue on Law and Legitimacy in Transnational Governance called "Comparing Forest Certification Schemes: The Case of Ratcheting Standards in the Forest

Sector (*Socio-Economic Review* 2010) and "The Experimental Turn in Environmental Sociology: Pragmatism and New Forms of Governance" (Springer 2010). Overdevest received her Ph.D. from the University of Wisconsin-Madison and her M.A. and B.A. from the University of Georgia.

Kevin E. Regan is an Associate Attorney in the Seattle office of Earthjustice, a nonprofit public-interest law firm. The views expressed in the chapter that Mr. Regan coauthored are his own and no endorsement by Earthjustice or its clients is intended. Previously, Regan was a trial attorney for the Natural Resources Section of the U.S. Department of Justice. In addition, he clerked for U.S. District Judge Jose E. Martinez in the Southern District of Florida and was an attorney for the U.S. Army Corps of Engineers in Chicago. Regan has published several articles related to environmental and natural resources law, including "The Need for a Comprehensive Approach to Protecting Rare Plants: Florida as a Case Study" (*Natural Resources Journal* 2004) and "Balancing Public Water Supply and Adverse Environmental Impacts under Florida Water Law: From Water Wars towards Adaptive Management" (*Florida Journal of Land Use and Environmental Law* 2003). Regan received his J.D. from the University of Florida and his B.S. from the University of Tennessee–Chattanooga.

Walter A. Rosenbaum is Professor Emeritus in political science at the University of Florida, where he served as the interim director of the Bob Graham Center for Public Service. He has recently been a visiting professor of environmental studies at the University of Michigan's School of Natural Resources and the Environment. Rosenbaum conducts research and writes extensively about issues related to environmental policy, energy policy, and risk management associated with environmental and public health issues. He is the author of the leading text *Environmental Politics and Policy* (7th ed. 2008). He is also the author of *Energy Politics and Public Policy* and numerous articles on environmental and energy policy. He recently completed an assessment of the environmental impact of the National Flood Insurance Program as a consultant to the Federal Emergency Management Agency and, earlier, another commission by the Center for Public Policy and Philanthropy at the University of Southern California. He has also served as a consultant to the U.S. Department of Energy, the South Florida (Everglades) Ecosystem Restoration Project, and the U.S. Environmental Protection Agency.

Sidney Shapiro is the Associate Dean for Research and Development and Distinguished Chair in Law at the Wake Forest University School of Law. Before joining the faculty at Wake Forest, Shapiro taught for twenty-five years at the University of Kansas School of Law. In addition, Shapiro was a trial attorney with the Federal Trade Commission and deputy legal counsel for the Department of Health, Education, and Welfare. He is a founding board member of the Center for Progressive Reform, a nonprofit research and educational organization of university-affiliated academics with expertise in the legal, economic, and scientific issues related to regulation of health, safety, and the environment. Shapiro teaches and writes about administrative procedure and regulatory policy. He is the coauthor of law school textbooks on administrative law and regulatory law and policy and has authored a one-volume administrative law treatise. His most recent coauthored book is *The People's Agents and the Battle to Protect the American Public: Special Interests, Government, and Threats to Health, Safety and the Environment* (University of Chicago 2010). Shapiro earned his law degree from the University of Pennsylvania and his B.S. from the University of Pennsylvania.

Rena Steinzor is Professor of Law at the University of Maryland School of Law and President of the Center for Progressive Reform. She teaches courses in administrative law, risk assessment, critical issues in law and science, and a survey of environmental law. Steinzor is the author of several books, including *Mother Earth and Uncle Sam: How Pollution and Hollow Government Hurt Our Kids* (University of Texas 2007); the coauthor (with Sidney Shapiro) of *The People's Agents and the Battle to Protect the American Public: Special Interests, Government, and Threats to Health, Safety and the Environment* (University of Chicago 2010); and coeditor (with Wendy Wagner) of *Rescuing Science from Politics: Regulation and the Distortion of Scientific Research* (2006). She has also written many articles on the topic of environmental regulation in the United States, including "Devolution and the Public Health" (*Harvard Environmental Law Review* 2000) and "The Legislation of Unintended Consequences" (*Duke Environmental Law and Policy Forum* 1998). Steinzor received her J.D. from Columbia University and her B.A. from the University of Wisconsin.

Joseph P. Tomain is a Dean Emeritus and Wilbert and Helen Ziegler Professor of Law at the University of Cincinnati College of Law. Tomain taught at Drake University School of Law and served as visiting professor at the University of Texas School of Law. He joined the University of Cincinnati law faculty in 1987 and served as dean of the College of Law for fifteen years. Before beginning a career in legal education, Tomain practiced law in New Jersey. He is chair of the Board of the Knowledge Works Education Foundation, the founder and principal of the Justice Institute for the Legal Profession, and a board member of the Greater Cincinnati Foundation. Tomain has published several books, including *Energy Law in a Nutshell* (West Group 2004), *Regulatory Law and Policy* (LexisNexis 3d ed. 2003), and *Energy Law and Policy for the 21st Century* (Rocky Mountain Mineral Law Institute 2000). Tomain received his J.D. from George Washington University and his B.A. from University of Notre Dame.

Preface

This book sets forth concrete proposals and ideas to guide the next generation of environmental law. The first generation of environmental law aimed to fully protect public health and the environment. It did so mostly through very detailed statutes and accompanying standards. This generation of law succeeded in meeting some of its goals and sparked significant progress toward meeting the rest. But it spawned an extraordinarily complex system that proved more difficult to implement than its creators had anticipated. Moreover, most of these statutes required that agencies prove harm before regulating, and many natural resource management statutes gave agencies broad discretion to balance competing values. Uncertainty and the broad discretion accorded agencies limited these statutes' success in achieving their stated goals.

We are nearing the end of a second generation of environmental law. This second generation carried out regulatory reforms ostensibly guided by a desire for economic efficiency. These reforms included greater reliance on cost-benefit analysis (CBA) to choose the goals of environmental law and market-based mechanisms as methods for achieving those goals. Although this approach enjoyed some successes, the CBA part of the agenda proved disastrous. By taking an insufficiently precautionary approach, the United States failed to act in a timely manner on global warming, which proved a much greater menace than economists and opponents of action had anticipated. CBA, while ostensibly aimed at rationalizing environmental law, usually simply provided a cover that allowed regulated polluters and ideologues favoring their interests to paralyze regulation.

The next generation of environmental law should build on the positive aspects of both of the previous generations of reform. It should embrace the precautionary approach of the first generation while embracing the second generation's goal of stimulating private markets to protect the environment. But we must address the shortcomings of the first generation and choose reforms that promise more vigorous protection of the environment, not less.

This book contains two parts. Part I focuses on a proposal to conserve public natural resources through the "National Environmental Legacy Act." Part II focuses on a proposal to stimulate positive technological changes through an "Environmental Competition Statute." Both parts explore related ideas and information to aid appreciation for how we can conserve our environment and stimulate appropriate technological progress. We see these two parts as complementary, since society needs both conservation of the good and progress toward something better.

Part I begins with a chapter describing its centerpiece proposal: the National Environmental Legacy Act (the Legacy Act). This statute would require us to define in concrete terms for the first time the environmental legacy we wish to leave to future generations and would provide a mechanism for systematically evaluating whether important decisions regarding public resources are consistent with preserving that legacy. At the same time, it would achieve the desirable goal of improving our ability to assess how our decisions are affecting our own future and whether they are consistent with our goals and priorities. As such, the statute would provide a tool to help us achieve the goal of sustainable use of our resources that is embodied in many of our current statutes but honored most often in the breach.

The Legacy Act concept draws on the lessons learned in more than thirty years of experience under the National Environmental Policy Act and the many other statutes by which we have sought to conserve both the quality and the quantity of our natural resources. A central lesson has emerged consistently from our experience under these statutes, and this lesson forms the cornerstone of the Legacy Act: we cannot reliably protect a natural resource legacy without a strong and enforceable substantive mandate. The chapter demonstrates how the Legacy Act would overcome the problems that have plagued our effort to achieve sustainable use of resources to date. The remaining chapters of Part I describe important ideas that undergird or support the Legacy Act concept or challenges that the Legacy Act must address.

Professor Shapiro focuses on administrative process and the key role it plays in our efforts to achieve sustainability. The chapter emphasizes the historical successes of citizen enforcement and describes three obstacles to enforcement by regulatory beneficiaries. Professor Shapiro describes the limitations that the courts' interpretations of standing and ripeness doctrines, as well as the courts' treatment of agency inaction, have imposed on that success. He then suggests legislative strategies to overcome these limitations in the design of the Legacy Act and argues that the Supreme Court should revisit and reverse recent and unwarranted interpretations of both standing and ripeness requirements.

Professor Ankersen and Kevin Regan write about the "shifting-baselines phenomenon," which both demonstrates the need for the Legacy Act and suggests a pitfall that it must help us to avoid. Humans' perception of normal or baseline environmental conditions can shift dramatically over time, and successive generations may not appreciate the degraded state of what they perceive as pristine and functional ecosystems. Ankersen and Regan explore the shifting-baselines phenomenon and the potential contours of the Legacy Act in terms of the ecologies of restoration, resilience, and reconciliation. They argue that the Legacy Act must encourage the restoration of already-degraded resources and ensure ecosystem resilience in the face of reasonably anticipated anthropogenic change. They also suggest that it is necessary to coordinate the Legacy Act's goals with broader landscape-level conservation efforts, including the accommodation of ecosystem functions, processes, and services on private lands. The chapter emphasizes the importance of restoring human ecological knowledge to prevent intergenerational ecological amnesia and to ensure the transfer of an environmental legacy.

Professors Brown and Angelo focus on an approach to analyzing the impact of decisions on natural ecosystems that could be applied as a central analytic tool to implement the Legacy Act's goals – emergy synthesis. Emergy synthesis is a quantitative method of valuation that relies on the intrinsic value of a resource or the services it provides. A core strength of the emergy approach is that, although it is quantitative, it does not require assigning an economic value to natural resources, which have both quantifiable and hard-to-quantify values. The latter are notoriously difficult to measure using standard economic techniques that rely primarily on determining consumer willingness to pay for the resource. Emergy analysis focuses on the energy embodied in the resource rather than trying to quantify the monetary value of the resource based on consumer preferences. Although emergy may not offer a complete picture of the value of the resources, it illustrates one promising technique that may play an important role in developing a more comprehensive method for assessing the value of natural resources to replace the outdated and inadequate cost-benefit framework. Thus, this chapter also illustrates more broadly the value of exploring some of the many newer approaches to analyzing environmental impacts and synthesizing available information.

Professors Overdevest and Mayer's chapter offers suggestions on how to design the information collection and dissemination components of the Legacy Act. They use a case study of bucket brigades – community-based air-quality monitoring and diffusion organizations that have developed across the country – to illustrate how the power of information is enhanced by creating incentives that engage both government and civil society in the process of collecting and sharing information. The case study suggests how engaging nongovernmental organizations and citizens with government oversight and monitoring can enhance both the information available and accountability. The authors' analysis suggests that an approach that collects and diffuses information at multiple administrative levels is more adaptive and functions more robustly in a climate where complexity and dynamism render complete information impractical. It also highlights areas in which further research is needed to fully understand the institutional dynamics.

Professor Rosenbaum also looks at information dissemination and offers recommendations on how the Legacy Act could ensure that the most effective use is made of information collected. The chapter begins by discussing three issues in the design and implementation of the Legacy Act that are of fundamental importance to achieving this goal: the process used to establish substantive standards, the treatment of uncertainty, and the selection of environmental indicators for baselines and monitoring. Professor Rosenbaum emphasizes that data collected under the Legacy Act must have policy relevance and be publicly accessible. The chapter also explores how the Legacy Act can both inform conservation efforts at the state and local levels and how those designing the Act can learn from these other levels of government. Professor Rosenbaum concludes by articulating the double standard of scientific appropriateness and policy relevance that information generated under the Legacy Act must meet.

Professor Steinzor argues for establishing a broader constitutional foundation for federal conservation efforts and details how a broader foundation may enhance the credibility and respect accorded a statute such as the Legacy Act. Although a broader constitutional foundation may not be essential to the Legacy Act, it may permit the statute to address a broader range of activities. The Property Clause of the Constitution provides a firm basis for regulating activities affecting the many natural resources found on federal lands, but this chapter explores the basis for conserving the quality of resources, such as air, water, wetlands, and endangered species that cannot be protected solely by regulating activities on publicly owned lands. The chapter outlines the impact of relying on the Commerce Clause as the sole basis for federal environmental protection. Professor Steinzor traces the evolution of Commerce Clause jurisprudence and its implications for environmental protection. She then traces constitutional foundations of the federal power to protect the public health, safety, and welfare. On the basis of this history, Professor Steinzor presents the core argument for relying on the Constitution's General Welfare Clause as affirmative constitutional authority for environmental, health, and safety legislation.

Part II begins with a discussion of its centerpiece proposal, an Environmental Competition Statute. This proposal seeks to promote competition among firms to maximize environmental protection. It seeks to emulate the dynamics of continuous improvement that characterize some highly competitive markets rather than the hypothetical efficiency that economists attribute to free markets to facilitate modeling exercises. It also seeks to relieve regulators of the responsibility to determine the limits of feasible protection by using the best performance from polluters as the benchmark others must meet to escape liability. The timidity of regulators has usually limited the efficacy of both traditional regulation and emissions trading approaches. This timidity also limits the efficacy of pollution taxes, as government officials must choose tax rates. The Environmental Competition Statute seeks to engage the creativity of the private sector directly rather than rely on government officials to set standards and tax rates.

An Environmental Competition Statute accomplishes this in a fairly simple way. It provides that anybody who makes an environmental improvement may collect the cost of making this improvement from competitors that pollute more, plus a premium preset by statute. This sets up a competition, a race to the top, in which those who perform best collect money and those who perform worse pay their cleaner competitors. This mimics the dynamics of markets, in which firms that innovate may capture market share from competitors so that they, in effect, take money from their less innovative competitors. This approach makes the achievements of the most environmentally capable firms rather than the timid actions of government bureaucrats the driver of environmental improvement. The discussion of the Environmental Competition Statute also explains why emissions trading and pollution taxes have proved fairly weak stimulators of environmental innovation and why such innovation is important. The remaining chapters provide additional justifications for the Environmental Competition Statute and champion some policies that will complement it in the next generation of environmental law.

Professor Adelman discusses a key premise of the Environmental Competition Statute – promoting technological innovation. The Environmental Competition Statute, of course, is premised on the notion that the private sector has great unused capacity to advance environmental technology, and it seeks to use competition to improve the environment as a means of bringing that capacity to the fore. Professor Adelman explores the level of government at which an Environmental Competition Statute can be most effectively implemented. He demonstrates that promoting innovation is a distinct regulatory end that is itself subject to a market failure – technology spillovers. Using efforts to address climate change as an illustration, Professor Adelman shows how failure to recognize technological innovation as a distinct regulatory end often leads policy makers to overlook the value of state and local regulation to complement federal regulatory efforts. The chapter then examines in detail the market dynamics that affect technological innovation and concludes that both state and federal governments can effectively implement a statute, such as an Environmental Competition Statute.

Professor Tomain offers a metaphor and vision for a transformation of the electric utility industry – the iUtility – and illustrates how an Environmental Competition Statute would be congruent with this vision. The chapter begins by describing today's electric utility industry, its history of regulation and deregulation, the critical function played by rate making, and the adverse consequences of this approach. Then Professor Tomain sets out a vision for a new regulatory compact based on the core ideas that animate the Environmental Competition Statute. He describes how a new approach to rate making and renewable portfolio standards can transform the industry and promote innovation.

In the concluding chapter, Professor Klein identifies a unified values framework that draws on values strongly associated with America's history and national character to support environmental sustainability and technological innovation. The chapter begins by describing the deep connection between patriotism and Americans' connection to the land. Professor Klein then surveys the evolution of environmental law, illustrating how both the first and second generations of environmental law embody the core American values of optimism, strength, and thrift. Building on this common foundation, Professor Klein dubs this framework "environmental patriotism." The chapter explores relevant metaphors associated with the values framework that can be used to support reforms that advance sustainability and examines how and why these metaphors are a critical tool in helping the public to understand the broad and profound implications of the decisions we make about the environment. Professor Klein details how the values framework and associated metaphors can help to educate the public about the meaning of both the Environmental Legacy Act and the Environmental Competition Statute, enrich public debate about the proposed statutes, and enlist a broad base of allies to support these two legislative initiatives.

We offer these two key reforms, the National Environmental Legacy Act and the Environmental Competition Statute, as cornerstones for the next generation of environmental law. They represent an attempt to grapple honestly with both the positive and the negative lessons from the first two generations of environmental law. We hope that these two ideas will receive support. But we also hope that they highlight the need for the next generation of environmental law to more comprehensively conserve our environmental legacy while simultaneously stimulating the deployment of technological advances that both promote conservation and meet our economic development needs. The analysis offered here supports these reforms but may stimulate further ideas. We welcome additional proposals on how the next generation can do better than the previous two in meeting these twin goals.

Acknowledgments

We are grateful to the many people who contributed to the development of these ideas and this book. The University of Florida School of Natural Resources and the Environment provided funding for the interdisciplinary roundtable, "The Next Generation of Environmental Law," at which the idea for this book was born. The Center for Progressive Reform provided a hospitable intellectual forum for the germination and growth of these ideas as well as invaluable staff support. Our contributing authors were engaged collaborators and helped us to develop and improve our proposed statutes through ongoing conversations and the writing and editing process. Alyson would like to thank Ryan Feinberg, Margaret Clune Giblin, Heather Halter, and Christina Storz, all of whom contributed materially to the development of the idea for the National Environmental Legacy Act at various points, and the University of Florida Levin College of Law and its Conservation Clinic for support of her research. David would like to thank Christy Ramsdell and the staff of Syracuse University's Law Library. Our special thanks go to our colleague, Professor Amy Sinden at Temple University's Beasley School of Law, for her careful reading of the entire manuscript and thoughtful comments on each chapter.

Alyson C. Flournoy
David M. Driesen

Part I NATIONAL ENVIRONMENTAL LEGACY ACT

1 The Case for the National Environmental Legacy Act

Alyson C. Flournoy

THERE IS VIRTUALLY UNIVERSAL AGREEMENT ACROSS THE political spectrum that we should protect the interests of our children and grandchildren in setting environmental, health, and safety policy. The concepts of sustainability and intergenerational equity, which advance this same objective, have become increasingly important in environmental law and policy debates in the past thirty years, both in the United States and internationally. In a large number of statutes, Congress and many state legislatures have embraced the goals of protecting a resource legacy for future generations and of promoting sustainable use of the nation's stock of natural resources.[1] In addition, in polls, the American public consistently expresses concern for how well we steward resources and has shown a strong recognition of a responsibility to future generations.[2]

Yet by any measure, it is clear that the United States is neither using its natural resources in a sustainable fashion nor systematically considering how today's patterns of resource use will affect the next generation. Report after report document the decline in supplies of fresh water, fish species and biodiversity, energy resources, and many of the values and services associated with those.[3] Many public natural resources are managed under statutes with notoriously open-ended standards that require federal agencies to

This chapter draws on a forthcoming report that also includes case studies illustrating how a "legacy act" would operate. Ryan Feinberg, Alyson C. Flournoy, Margaret Clune Giblin, Heather Halter & Christina Storz, The Future of Environmental Protection: The Case for a National Environmental Legacy Act: A Center for Progressive Reform Report, *available at* http://www.progressivereform.org/articles. The idea for a legacy act has developed over several years with contributions from all of the report's coauthors.

"balance" a variety of often-incompatible uses, many of which degrade or deplete relevant resources.[4] Many of these statutes contain no enforceable standard mandating protection of any particular quality or quantity of a resource. Instead, they generally charge the relevant agency with developing a plan for the resource that considers a list of competing potential uses. The agency is then granted considerable discretion to decide which of the competing uses to permit and on what terms. Not only do the statutes entrust the agencies with broad discretion to permit degradation and depletion of public natural resources, but also cases interpreting the statutes have removed some of the few effective checks on agency discretion.[5] Courts have limited both citizens' ability to enforce agency commitments to preserve resources set forth in agency management plans and citizens' ability to force agencies to take action to comply with affirmative statutory mandates if such duties are found not to be sufficiently "discrete."[6]

In light of these and other shortcomings of current law, the next generation of environmental laws should act on the laudable, widely embraced, but largely unrealized goal of protecting a resource legacy for future generations. This chapter outlines a proposal for a new statute – the National Environmental Legacy Act (Legacy Act) – that would require us for the first time to define in concrete terms the environmental legacy we wish to leave to future generations and provide a mechanism to ensure that we preserve that legacy.[7]

The Legacy Act proposed here focuses on preserving a public natural resource legacy for the next generation, because the quality and quantity of available natural resources are key determinants of the options and quality of life that future generations will enjoy. Public natural resources include water and land, as well as the ecosystems, biodiversity, and minerals found in or on the land or water. The Legacy Act seeks to protect both the resources and the many values and services they provide. As proposed, the Legacy Act applies to resources that are under public ownership or management or are protected by the public trust doctrine, but not to natural resources in wholly private ownership.

If we acknowledge that public natural resources are a significant form of natural wealth, concern for the resource legacy we leave follows naturally.[8] The concept of defining and preserving a resource legacy builds on basic principles of wealth management. Just as an estate plan enables individuals with private wealth to ensure that their wealth is protected for the next generation, the Legacy Act provides a mechanism to ensure that public wealth is preserved for the next generation and not depleted or spent today. Financial professionals universally recommend that individuals adopt a savings plan if they truly wish to save for the future; they do not recommend that they spend all their currently available income. A default position that savings will occur and that a specified amount will be saved for the future is the keystone of success.[9] The Legacy Act adopts this same approach to conserving natural resources. The statute requires us to determine a threshold of resources that we commit to leave to future generations, and it ensures that necessary savings will occur. Just as a savings plan requires that one protect and set aside the money needed for savings and spend only what income remains, the Legacy Act requires stewards of public resources to set aside a defined level of resources for future generations and to spend or use only that which remains or can renew itself.

The central and innovative feature of the Legacy Act is a clear and enforceable conservation mandate that focuses on sustainability and constrains resource use.[10] Whatever standard of sustainability is chosen, the key is that a legacy is defined and a clear and enforceable mandate that will preserve the desired legacy is imposed.[11] The importance of a clear, enforceable mandate is perhaps best illustrated by considering the starkest alternative to adopting such a mandate – that is, to impose no enforceable standard, to make no meaningful commitment on what our legacy will be. Such a choice would reflect what is called a "spend-down ethic." It would reflect the decision to use resources according to the dictates of our current short-term needs, however high the costs we are leaving our children and however strong the likelihood is that they will experience dislocation and loss from the anticipated depletion of key public natural resources, including fossil fuels, freshwater supplies, fisheries, and so on. A spend-down ethic is premised on a lack of any ethical or moral commitment. Such an approach leaves our children and their children to overcome these challenges and to use as they see fit whatever resources may remain.

Although it may sound harsh, this spend-down ethic is in fact the legacy we are currently pursuing. We do not embrace it openly, but by relying on highly discretionary resource management statutes with open-ended balancing tests, purely procedural impact assessment under the National Environmental Policy Act (NEPA), and analytic approaches like cost-benefit analysis that fail to adequately account for the future, we are tacitly accepting that we do not know and do not care what the impact of our resource use will be. Blithe and unfounded speculation and assumptions about the greater wealth and technological prospects of future generations are routinely offered as the justification for excluding consideration of our children's interests from critical decisions or for discounting them dramatically.[12]

If the Legacy Act is to avoid perpetuating this same ethic – in other words, if it is to achieve some purpose – it must include a clearly articulated mandate for conserving some defined quality and quantity of key public natural resources and a related prohibition on any resource use that will impermissibly deplete or degrade these legacy resources. Absent such an enforceable prohibition, agency decision makers will be subjected to the same pressures to which they respond under current law. These pressures have eviscerated the sincere but ineffectual aspirations to sustainability embodied in statutes like the Multiple-Use Sustained Yield Act,[13] the Magnuson-Stevens Fishery Conservation and Management Act,[14] and the National Environmental Policy Act.[15]

Notwithstanding the critical importance of an enforceable mandate for sustaining a defined quality and quantity of individual resources, the statute must also take account of two fundamental insights from ecology that arguably stand in tension with the goal of preserving a defined quality and quantity of individual resources. The first of these ecological insights is that, to preserve natural resources, we must ensure not just the continued existence of the individual resources but also the health of the ecosystems of which they are part. The solution embedded in the Legacy Act is to require that, while pursuing the mandate to prevent impermissible depletion and degradation of protected resources, resource stewardship agencies also plan to ensure that the ecosystems of which the resources form part maintain their resilience.[16]

The second ecological insight arguably in tension with the Legacy Act's central feature is that natural systems are dynamic and change even in the absence of human intervention. The Legacy Act seeks to respond to this insight but without reverting to a statutory design that simply accords agencies standardless discretion to manage adaptively, thereby replicating the shortcomings of current laws. The solution incorporated in the Legacy Act is to set a clear standard and focus regulatory efforts on preventing changes in natural systems that result from postenactment human conduct.[17]

Given the evidence of a long-term trend of depletion and degradation of natural resources, as well as the systemic biases in favor of economic and consumptive use of resources, we need a better method to help us account for impacts on future generations. The Legacy Act would help us to make conscious policy choices about the legacy we leave and would provide the tools to ensure that we manage our natural resource wealth wisely and in accord with our chosen priorities.

This chapter offers a brief overview of the contours of the proposed statute. The remainder of this chapter discusses some of the key attributes of the proposed statute, identifies some issues that the design of the statute raises that warrant further debate, and addresses some of the objections to the Legacy Act concept that will likely be raised.

The Contours of the National Environmental Legacy Act

The concept of the National Environmental Legacy Act is to define and protect a legacy of public natural resources for future generations, something no statute has done successfully to date. Building on the goals already expressed in numerous laws, the Legacy Act would, for the first time, require management of public resources to conserve some stock of resources for future generations. Embrace of the Legacy Act concept would impel us to identify our long-term goals and would then help us chart and maintain a course to achieve our shared goals. It would also improve our decisions over the long term by generating the information base needed to support adaptive learning.

For purposes of discussion, I propose a very broad definition of public natural resources that includes all resources under federal ownership or protected by the federal public trust doctrine, together with all the values and services associated with those resources. Thus, public natural resources would include forests, wetlands, uplands, and all other types of ecosystems found on public lands and all the species of life found in these ecosystems, as well as fisheries under federal protection or control. Minerals encompassing an array of hard-rock minerals as well as oil, gas, and other nonrenewable energy sources would also be covered. The values and services these resources provide to humans are numerous and varied. For example, lands within a national forest may provide timber for consumptive use; habitat for wildlife; carbon sequestration; watershed and erosion protection; and aesthetic, spiritual, and recreational values, to name a few.[18]

At a minimum, the idea of the Legacy Act envisions a statute that defines the public natural resource legacy we wish to preserve and prohibits all actions that will degrade or deplete the defined legacy. These two core objectives of the statute are guideposts that suggest the general contours of the statute. Building on these objectives, I propose the following model to achieve the goals of the statute.[19] A more detailed discussion of its key provisions follows this general description of the statute.

Section 1 – Goals and Policy: The statute should set out the goal of defining and preserving a legacy of public natural resources for present and future generations of Americans. The statement of goals and policy should also describe in affirmative terms the legacy we wish to leave, defined in relation to our existing stock of resources.

Section 2 – Designation of a Legacy Period: The statute should designate a fixed period of years that constitutes the legacy period over which public natural resources must be conserved.[20]

Section 3 – Prohibited Degradation or Depletion of Legacy Resources: The statute should set forth in clear and enforceable terms the maximum level of degradation or depletion of resources, if any, that will be permitted over the course of the legacy period. This is critical to ensure the enforceability of the statute. The statute should include distinct standards of permissible degradation and depletion for renewable and nonrenewable resources. The statute should in broad terms prohibit actions by any person[21] that may produce impermissible degradation or depletion of a legacy resource during the legacy period. Section 3 by its terms can effectively only constrain postenactment human conduct. In other words, its prohibitions will not preclude changes that are caused solely by (1) human action taken prior to the date of enactment of the Legacy Act or (2) changes in resources or ecosystems that are not caused by human actions.

Section 4 – Designation of Legacy Resource Stewardship Agencies: The statute should designate an existing federal agency to serve as the resource stewardship agency for each public natural resource.[22]

Section 5 – Development of Metrics and Collection of Baseline Data on Resource Quality and Quantity: Each stewardship agency should be charged with developing the implementing regulations that designate appropriate metrics of quality and quantity for the resources for which they are stewards. The statute should both mandate and authorize adequate funding for the collection of baseline data on the quality and quantity of all public natural resources employing these metrics.

Section 6 – Promulgation of Rules Defining Maximum Permitted Levels of Degradation and Depletion over the Legacy Period: Each stewardship agency should be required to promulgate rules that translate the substantive prohibition articulated in Section 3 into enforceable standards for each relevant resource, expressed in terms of the metrics developed under Section 5. In addition, each stewardship agency should be directed to identify tipping points for each ecosystem under its stewardship. By "tipping point," I mean resilience thresholds – thresholds beyond which degradation of resource quality or quantity will cause loss of ecosystem resilience.[23]

Section 7 – Stewardship Agency Mandate to Ensure No Impermissible Degradation Will Occur: The statute should limit stewardship agencies' discretion under existing law by requiring that each stewardship agency ensure that no degradation or depletion in excess of permissible limits will occur during the legacy period. The statute should also specifically mandate that each stewardship agency develop a legacy "plan" to demonstrate how it will ensure that the mandated resource legacy is conserved over the legacy period and conform its actions to the legacy plan.[24]

Although the Act's prohibitions constrain only post–Legacy Act human conduct, agencies should be required to monitor, assess, and consider degradation and depletion from all sources in planning and in making decisions that affect legacy resources. The statute should also mandate that each stewardship agency ensure the resilience of relevant ecosystems that encompass legacy resources. Where the duty to prevent degradation or depletion of a specific resource and the duty to ensure resilience of the relevant ecosystem conflict, the duty to maintain resilience should prevail.

Section 8 – Ecosystem Resilience Assessment: Should preenactment human conduct or non-human-induced changes cause significant degradation or depletion to a legacy resource, the agency should be required to perform a resilience assessment to determine whether the ecosystem retains the capacity to persist in light of the degradation or depletion. If the ecosystem can persist, the agency should be required to modify its legacy plan as needed to promote the continued resilience of the system. In cases where the system can no longer persist and has flipped to a new behavior regime, the question of whether to attempt to restore the system or to seek to promote the resilience of the new regime should be addressed by the stewardship agency following public comment. Factors to be considered would include the values and services provided by the ecosystem in its prior state and in its current state, the uniqueness of the resources, and the cost of restoring the ecosystem to its prior behavior regime.[25]

Section 9 – Enforcement: To ensure enforcement, both the stewardship agency and citizens should be granted enforcement authority. A citizen suit provision with fee shifting would be a critical component of the statute.[26] This provision should authorize any person to bring an action to enjoin and seek penalties for any action that impermissibly degrades or depletes public natural resources. The statute should also permit citizen suits against any stewardship agency to enforce other agency duties under the statute, including the duty to collect information, the duty to develop or update a legacy plan, and the duty to conform agency actions to the terms of the legacy plan.

Section 10 – Monitoring and Adaptive Learning: The statute should require and authorize funding for ongoing monitoring of legacy resources and the ecosystems of which they form part and should require stewardship agencies to update legacy plans according to a fixed schedule.

Section 11 – Exceptions: The statute should allow for a narrow exception to its prohibition on degradation or depletion in two circumstances: (1) if it can be shown by clear and convincing evidence that foreseeable technological advances or the availability of substitute resources will obviate the need for and value of the resource in question or (2) if impermissible degradation or depletion is clearly in the public interest, no acceptable alternative that will not cause impermissible degradation or depletion exists that will serve the public interest adequately, and the impacts to all services and values to be impaired can and will be mitigated.[27]

In the following sections, some of the key attributes of the Legacy Act are discussed. In addition to explaining the proposed statute in greater detail, this discussion identifies issues on which further work is needed to permit full development of a statute.

Key Attributes of a Legacy Act

1. Interplay of the Proposed National Environmental Legacy Act and the National Environmental Policy Act[28]

The Legacy Act follows a general approach first made popular with the National Environmental Policy Act: a single statute that applies across the board to decisions affecting a wide range of resources. This approach supplements the resource-by-resource approach that characterizes many of our public natural resource conservation and management laws. Although NEPA has been criticized as generating unnecessary paperwork and inadequate results,[29] it has also been recognized as having achieved significant gains.[30] The success of NEPA's innovation is demonstrated by its widespread imitation around the world and by some U.S. states.[31]

In addition to sharing with NEPA a scope that encompasses actions affecting a wide variety of federal lands and resources, the Legacy Act also echoes NEPA in its embrace of a policy in favor of preserving resources for future generations. The National Environmental Policy Act is premised on a recognition of "the profound impact of man's activity on the interrelations of all components of the natural environment, particularly the profound influences of population growth, high-density urbanization, industrial expansion, resource exploitation, and new and expanding technological advances."[32] In its statement of policy, NEPA recognizes "the critical importance of restoring and maintaining environmental quality to the overall welfare and development of man."[33] It aspires to create and maintain conditions under which "man and nature can exist in productive harmony, and fulfill the social, economic, and other requirements of present and future generations of Americans."[34] In seeking to carry out this ambition, NEPA adds to the federal government's responsibilities a duty to use its resources "to the end that the Nation may – (1) fulfill the responsibilities of each generation as trustee of the environment for succeeding generations."[35] Although NEPA also requires agencies to seek to "attain the widest range of beneficial uses of the environment," this is to be "without degradation, risk to health or safety, or other undesirable and unintended consequences."[36] Thus, NEPA can be fairly said to have at its core a legacy ambition – a conscious recognition of a duty we owe as a nation to future generations and a policy that adds to the mandate of federal agencies a duty to take steps to fulfill the role of steward of our natural resources.

From reading these statements of policy, one might think the proposed Legacy Act redundant. In practice, however, it is widely recognized that NEPA has failed to fulfill many of these lofty ambitions.[37] Consequently, the Legacy Act's design seeks to achieve these goals through very different means. Thus, the Legacy Act prohibits resource degradation or depletion that exceeds a specified standard. In other words, it contains a clear enforceable standard, not merely a general statement of policy backed by procedural requirements.[38] Unlike NEPA, the Legacy Act is also designed to require an assessment of the resource legacy we will leave the next generation – not just to evaluate the impacts we anticipate from a particular proposed action but also to consider specifically the position in which those effects will leave our children and their children.

Also, unlike NEPA, the Legacy Act is designed to permit adaptive learning.[39] To ensure that we can learn from our experience, it would require ongoing monitoring of resource conditions. This would in turn provide data to help us make informed decisions in the future and to assess whether our past analyses were accurate.

NEPA has also been criticized as failing to generate needed information on cumulative impacts. The Legacy Act responds to this by replacing NEPA's focus on individual actions with a focus on the cumulative effect of actions on a particular resource. Instead of generating information on anticipated impacts of individual actions, the Legacy Act will define the quantity and quality of resources we wish to leave at the end of the legacy period and prohibit incursions into that legacy.

Notwithstanding its significant shortcomings, NEPA has been rightly and widely praised for its ability to bring the environmental consequences of proposed actions to public attention and for educating and focusing government agencies on the environmental effects of their decisions, impacts that would otherwise have been overlooked. There are many cases in which this has affected the decisions ultimately made.[40] Thus the Legacy Act seeks not to replace NEPA but to complement it.

NEPA's most powerful and important mandate requires that federal agencies assess the environmental impacts of their actions by preparing an environmental impact statement (EIS) for every major federal action that significantly affects the human environment.[41] Yet it lacks a substantive standard to govern decisions. Thus, although NEPA has proved a valuable tool to ensure that federal agencies consider environmental consequences of federal actions, it offers no guarantee that natural resources will be protected today, not to mention for future generations.[42]

This is captured by the common complaint that NEPA "lacks teeth."[43] Federal agencies complying with NEPA must prepare required documentation[44] that details impacts of the proposed action and alternatives to the proposed action, but there is no requirement that the agency then pursue the most environmentally preferable alternative identified in the analysis.[45] Because NEPA's main enforceable requirement is that the government's documentation of impacts and alternatives comply with NEPA's standards for completeness, agencies labor to ensure that documentation details every impact and appropriate alternative. Yet NEPA does not ensure that the information on environmental impacts actually affects the agency's decision. Under NEPA, even if agencies comply completely with NEPA, critical public natural resources can still be completely depleted and degraded. For this reason, the Legacy Act departs substantially from the impact assessment model of NEPA and imposes an enforceable substantive mandate.

2. An Enforceable Substantive Mandate (Sections 1 and 3)

To merit the name "Legacy Act," the statute must have teeth: it must include a strong, substantive mandate that protects resources instead of merely requiring an assessment of impacts. In the Legacy Act, Congress would articulate the enforceable substantive mandate in two different forms. First, in section 1 of the statute, Congress would describe in narrative terms the public natural resource legacy we commit to leave to future generations, defined in relation to the current stock of public natural resources. For example, Congress might determine that the legacy should be an identical stock of resources in terms of quantity and quality. Or it might determine that the legacy should be a stock of resources that is not "substantially diminished in quality or quantity." Or it might choose some other standards of quality and quantity of resources that we commit to preserve for future generations. This is the affirmative vision of the legacy and is central to the Act.

In section 3 of the statute, Congress would then translate this affirmative vision into a clear and enforceable prohibition that applies to all activities affecting public natural resources. So, for example, if in section 1, Congress defined the legacy to be preserved as "*identical* in quantity and quality to the existing stock of resources," then section 3 would impose a corresponding prohibition on any action that may degrade or deplete public natural resources at the end of the legacy period, with perhaps a de minimis exception. In contrast, if the affirmative vision embraced in section 1 is a legacy of resources that are not "*substantially* diminished in quality or quantity" from those we have today, then section 3 would prohibit actions that may "*substantially* diminish the quality or quantity" of public natural resources over the course of the legacy period.

Whatever standard is chosen, it is important that the standard be articulated by Congress in clear and operational terms. To the extent possible, the statute should provide clear guidance on the quantitative and qualitative measures of depletion and degradation of resources that will be permitted.[46] Because there has not been meaningful public debate on the precise contours of the legacy that we as a society want to leave our children, it would be premature to assume the precise value choices that such a statute should embody. However, public debate surrounding the enactment of the Legacy Act would provide an important occasion to crystallize public sentiment on precisely what we want our legacy to be and to resolve this question. The statute would then be drafted to ensure that we preserve that legacy. If our vision of the appropriate legacy subsequently changed, the law could be amended to reflect that change.

In light of the broad public support for the concept that we should leave our children at least as well off as we ourselves are, debate might appropriately begin with a presumptive standard of sustainable use for renewable public natural resources.[47] Sustainability's most basic justification is intergenerational equity: a desire that present development not compromise the ability of future generations to meet their needs.[48] Inherent in a commitment to sustainability is the requirement that goals for economic development, social development, peace and security, and natural resources protection should be met for both present and future generations.[49]

This chapter proposes one standard as a starting point for discussion, mindful that the standard represents a particularly important public value choice and without the intent to preempt that choice. Therefore, I propose a standard that prohibits any degradation or depletion of covered renewable resources over the legacy period. For nonrenewable resources, selecting a standard would involve more complex trade-offs. I propose a standard that prohibits significant degradation or depletion of nonrenewable resources, a standard that would require considerable further elaboration. But the idea would be to allow some drawdown of these resources, perhaps an annual percentage of permissible depletion. These standards seek to balance the needs of the present and future generations. Therefore, the standard for nonrenewable resources is less restrictive than the strict standard of sustainable use that applies to renewable resources, but both leave a defined legacy of public natural resources.[50]

3. Baseline Information and Metrics (Section 5)

As envisioned, the Legacy Act depends on our ability to assess the quality and quantity of the existing stock of public natural resources now and in the future. Therefore, adequate provision for the collection of baseline information and ongoing monitoring is critical to the Act's success. However, an open-ended demand for information on the quality and quantity of public natural resources could prove an endless and excessively costly quest.

Therefore, drawing on an idea developed by Professors Shapiro and Steinzor in another context, I propose the use of metrics to serve as a shorthand for assessing the quantity and quality of resources.[51] Shapiro and Steinzor define metrics as short and concise measures, selected by an independent body of experts, addressing the most important issues, and focusing on outcome rather than output.[52] The metrics would be selected to assess the status of resource quantity and quality in place of comprehensive information. Metrics would provide guidance on the baseline information to be collected as well as define the parameters for ongoing monitoring of public resource quantity and quality. Notwithstanding the challenge that developing such metrics presents, there is a wealth of expertise that agencies have developed in implementing NEPA and resource management and protection statutes, as well as much academic expertise, that could be brought to bear on the task.

An example of a metric that is sometimes used to assess ecosystem health is the status of an indicator species, a species whose status provides information on the overall health of the ecosystem and of other species in that ecosystem. An analysis of the range of possible metrics and their suitability for achieving the purposes of the Legacy Act is beyond the scope of this chapter. To ensure that the best available scientific and economic methods are employed in developing metrics and in resource planning, Congress should include a mandate to convene an interdisciplinary committee of experts to help develop the metrics, following the model of the Committee of Scientists, which was created to develop regulations under the National Forest Management Act, or a broader undertaking modeled after the Intergovernmental Panel on Climate Change or the Millennium Ecosystem Assessment. Under either model, the goal would be to bring leading scientists together to assess the state of knowledge and to determine whether consensus exists on the state of our information and analytic methods and how best to assess the quantity and quality of various resources, given the state of knowledge and available analytic methods.[53] Such a committee could be reconstituted every five to ten years to revisit the methodologies and information demands under the Legacy Act, to ensure that the most up-to-date techniques are employed.

In addition to developing metrics for assessing the quality and quantity of legacy resources, agencies will need to develop measures to assess the resilience of relevant ecosystems and to collect relevant data. This data would provide stewardship agencies with information on the range of conditions in which the resources, values, and services associated with the ecosystem will persist. Ecologists suggest that a focus on resilience has significant advantages not provided by a focus on individual resources.[54]

The information demands of the Legacy Act should also be designed with an eye toward maximizing the transparency and usefulness of the information. Thus, drafters should seek to structure information demands and analysis so that state and local decision makers, as well as interested advocates for the public interest, could benefit from the information generated.[55] To maximize transparency, the statute should make the information easily and broadly accessible. For example, the statute could mandate that baseline information and updated monitoring data be made accessible on the stewardship agency's Web site. Ensuring that the data generated under the Legacy Act is publicly available and in a form that is readily usable would promote transparency and could enhance the quality of decision making not just by federal agencies but by state and local governments and private enterprise as well.[56]

Whatever information demands and analytic techniques are ultimately adopted, assessment and monitoring under the Legacy Act would entail costs. However, those costs should not be viewed as waste. The monitoring costs associated with the Legacy Act would enable the type of adaptive learning that many environmental scholars have recommended, thus enabling us to make better and better decisions as a society over time as our base of data grows. Indeed this information collection and analysis could very likely prove less costly than the far less helpful but prevalent information demands imposed by cost-benefit analysis.[57]

4. Ecosystem Resilience: The Bottom Line[58]

To avoid an overly narrow focus on individual resources that ignores the importance of scale and the dynamic nature of ecosystems, sections 6, 7, and 8 all incorporate the concept of ecosystem resilience as an adjunct to protection of specific legacy resources. C. S. Holling developed the concept of ecological resilience as a way to describe the "persistence of relationships within a system and . . . a measure of the ability of these systems to absorb changes" and still persist.[59] Thus, it can help us to describe the degree of disturbance a system can tolerate before it flips into another behavior regime.[60] Resilience expresses the ability of a system to rebound from disturbance and the point at which a disturbance triggers a shift in the structure of the system.[61] The concept of resilience enables us to step back from the realities of change, uncertainty, and the dynamic nature of ecosystems to focus on what might be called the degree of play in a given system – how the system responds to perturbation.

Section 6 incorporates this concept by mandating that stewardship agencies develop measures for assessing the resilience of the relevant ecosystems, and section 7 demands that stewardship agencies maintain the resilience of the relevant ecosystems. This mandate ensures that agencies will keep their eyes on the big picture – the health of the ecosystem – as well as individual resources, services, and values that the ecosystem provides. As we have learned from experience under the Endangered Species Act, ultimately it is futile to seek to preserve a particular resource without preserving the ecosystem on which it depends.

Some may see the demand for information under these provisions as an obstacle. However, arming resource stewards with information on critical thresholds and thus the range within which they must manage the components of the relevant ecosystem may simplify management and make it far more efficient. Indeed, a leading ecologist has suggested that resources would be better directed to assessing resilience, adaptive capacity, and renewal capability of ecosystems rather than "variables of the moment" like water levels and population numbers.[62] The Legacy Act therefore focuses attention on identifying key metrics, including measures of resilience, rather than seeking to collect complete information about every aspect of every resource. Determining how to implement a resilience mandate and whether it is feasible is precisely the type of question that a committee of experts should be charged with addressing.

Section 8 draws on the concept of resilience to address the problem of impacts to resources that either are not the result of human conduct or result from preenactment human conduct. The ongoing and anticipated effects of climate change vividly illustrate why the Legacy Act's design must account for these sources of resource degradation – because they cannot be prevented by prospective regulation. The Act addresses these types of change in several ways. First, it requires that stewardship agencies monitor and collect data on resources without regard to the cause of the change. Second, it requires that the agency's legacy planning consider and account for all types of change. In other words, the agency must still strive to maintain the legacy notwithstanding these types of change.[63]

However, where significant degradation occurs that is not the result of human conduct or that is the result of preenactment conduct, the statute recognizes that it may not be possible to retain the legacy that we sought to preserve. The predictions of migrating ecosystems, coastal erosion, and flooding as a result of climate change make clear that we must be prepared for circumstances in which it no longer makes sense to seek to preserve resources in the location where they once were found.

Therefore, in cases where forces beyond the reach of the Legacy Act cause significant degradation or depletion of resources, the agency is directed to perform a resilience assessment on the relevant ecosystem. This will permit a reasoned assessment of the likelihood that the ecosystem can persist in the face of the changes. If the assessment suggests that the ecosystem cannot persist – that it has flipped or inevitably will flip into a new behavior regime without intervention – the statute does not assume that restoration should always be undertaken.

For example, if the change to the ecosystem results from a permanent shift in the climate, such restoration might be both costly and futile. In other cases, the change might be reversible and the ecosystem might be able to be restored, and its values and services might warrant the cost to do so. Because the decision on whether to seek to restore an ecosystem in these circumstances involves important value choices and technical decisions, the statute should require that such a decision be the subject of at least a notice and comment proceeding to permit public input and debate of the value choices and technical questions. A decision to undertake restoration would be consonant with the mandate under section 7 of the Act. Should a decision be made to allow the ecosystem to shift to a new behavior regime, section 8 mandates that the agency update its legacy plan to reflect this change.

5. The Role of Stewardship Agencies (Sections 4, 5, 6, 7, and 9)

The Legacy Act seeks to build on the expertise and existing administrative structure of resource management agencies in the federal government. Thus, it envisions designation of a single agency to serve as the "stewardship agency" for each resource covered by the Act and for all the values and services associated with the relevant resource. The Act envisions a number of duties that accompany designation as a stewardship agency:

- Promulgation of rules designating metrics of resource quality and quantity
- Collection of baseline information about the quality and quantity of the resource and ecosystem resilience
- Ongoing information collection
- Promulgation of rules defining what constitutes impermissible degradation or depletion of the relevant resource in terms of the selected metrics
- Development and periodic updating of a plan that ensures that the legacy resource will not be impermissibly depleted or degraded over the legacy period
- Compliance with the plan
- Avoiding and preventing actions that impermissibly degrade or deplete the resource

In addition to these duties, the statute also provides the stewardship agency and citizens with enforcement authority and responsibilities. Thus, the stewardship agencies play a central role in implementing the Act and ensuring preservation of the resource legacy. Some of the duties imposed under the Legacy Act may complement an agency's duties under existing law. Others would directly conflict with and therefore modify the agency's existing duties. The most important instances in which the Legacy Act duties would conflict with and should modify and supersede agency duties under existing statutes are the provisions that prohibit the agency from impermissibly depleting or degrading the resource (section 3) and that require the agency to conform its actions to the legacy plan it develops (section 7). For the Legacy Act to be effective, these mandates must override agency discretion afforded under other resource management laws, acting as a check on agency actions and inaction that would cause impermissible depletion or degradation of the resource.

6. Coping with Uncertainty (Sections 3, 7, 8, and 10)

The obstacles that uncertainty creates for protecting natural resources, health, and safety under existing statutes are well documented.[64] Many environmental regulatory statutes require that an agency provide a specified measure of proof of harm before regulatory constraints are imposed. Where the burden to prove harm falls on the agency seeking to regulate potentially harmful actions, the inevitability of incomplete and uncertain information can act as a significant impediment to regulation.[65] The difficulty our society faces in coping with scientific uncertainty and the value of a precautionary approach are manifest in our experience with developing a policy response to global warming. Policy development was delayed by a prolonged period in which the views of a few scientists on the margin of the scientific community were cited as sufficient evidence of uncertainty about the cause of global warming to make delay the only reasonable course of action.[66]

Given the purposes of the Legacy Act and our experience under existing law, it is essential that the Legacy Act's prohibition be framed to adopt a precautionary approach.[67] Thus, the statute should place the risk created by uncertainty on the party that wishes to deplete or degrade resources. All doubts should be resolved in favor of preserving the resource. Although the precautionary principle in its broadest form has met with significant criticism,[68] shifting the burden of proof to achieve environmental protection is a limited form of precaution that does not raise all of the same concerns. Therefore, section 3 of the statute should be framed to prohibit any person from taking any action that may cause impermissible degradation or depletion of any legacy resource, or it should employ similar or even more precautionary language. In contrast, in granting an exception pursuant to section 10, doubt should be resolved against granting the exception. The statute should also make explicit that any doubt created by inadequate or uncertain information under any section of the statute should be resolved in favor of protecting the legacy, because uncertainty will frequently limit our ability to predict impacts.

In addition to embracing a precautionary approach, the statute should include a transition provision, imposing a partial or complete moratorium on significant activities with potential effects on legacy resources pending collection by stewardship agencies of mandatory baseline data and the development of implementing rules.[69] Such a provision would align the interests of resource users and the public to support development of baseline information and thereby reduce uncertainty. Under the usual regulatory model, resource users have incentive to delay and oppose agencies in their quest to collect necessary information and develop rules, because until these are in place, the resource can be used freely.[70] A transition provision would reverse this incentive structure, creating an incentive for private parties to share relevant information with the agencies and to support prompt promulgation of agency rules implementing the statute rather than to favor delay.[71] As with the experience under Proposition 65 and the European Union's REACH program of chemical regulation that Professor Karkkainen has described, this would create an incentive to generate information and support regulation that typical regulatory statutes lack.[72]

7. Exceptions

The Legacy Act should allow for the possibility that the public interest will be better served in some instances by degrading or depleting a resource today where mitigation can be and is undertaken to offset the lost values.[73] Therefore, as described earlier, the statute should allow for a narrow exception to its prohibition on degradation or depletion in two narrow circumstances. An exception to the prohibition should be allowed if it can be shown by clear and convincing evidence that (1) foreseeable technological advances or the availability of substitute resources will obviate the need for or value of the resource in question; or (2) the action that will cause impermissible degradation or depletion is clearly in the public interest, no acceptable alternative that will not cause impermissible degradation or depletion exists that will serve the public interest adequately, and the impacts to the services and values to be impaired can and will be mitigated.

This provision allowing exceptions could operate in two different contexts under the statute. First, it might apply in the context of an agency's duty to develop rules under section 6 of the statute that ensure no impermissible depletion or degradation of a particular resource. An agency could trigger the exception provision by demonstrating in the rule-making record by clear and convincing evidence that either prong of the exception is met as to a particular resource or value. If it did so, the statutory mandate for preserving the resource or value would be modified to the extent the evidence warrants and the agency could lawfully promulgate rules that deviated from the statutory mandate to that extent. In that case, agency rules that deviated from the statutory mandate in sections 1 and 3 would be in compliance with the Act under the exception.

The second context in which an exception might be available would be as a defense to an enforcement action. A person subject to enforcement for impermissible degradation or depletion could defend against enforcement by demonstrating by clear and convincing evidence that either prong of the exception is met as to the action giving rise to the enforcement.

Application of the exception to permit resource degradation or depletion should be extremely rare in cases in which the affected values include aesthetic, spiritual, or other hard-to-monetize values, or where endangered species or pristine areas would be affected. By definition, mitigation of such losses or creation of such values is extremely difficult, and substitutes are often not available for noneconomic services and values. Where the only values or services affected are of the sort frequently traded and valued in markets, the possibility of mitigation would be more realistic and thus an exception more likely to be allowed.

In seeking to claim an exception, the proponent would bear the burden of proof on all elements. Unlike under many pollution control statutes, and consistent with the precautionary principle, the burden of amassing adequate proof should work in favor of conservation instead of depletion or degradation. This burden should provide a sufficient disincentive to proponents that exceptions would be pursued only in cases in which the benefits were substantial and the degradation of resources limited to effects that are relatively easily assessed and mitigated.

8. Monitoring and Adaptive Learning

A major criticism of NEPA's use of information is the limited, onetime use of information to inform a single decision.[74] This limits the value of the information collected and precludes adaptive learning. To ensure that we can learn from the experience gained during the legacy period, the Legacy Act should require ongoing monitoring of all resources covered under the Act and regular updating of legacy plans by stewardship agencies. This would in turn provide data to help agencies and the public to make informed decisions in the future and to assess whether past analyses were accurate. Without this information, we lose the opportunity to improve decision making by observing the deviation between ex ante predictions and actual impacts.[75] Postdecision monitoring would also permit adaptive responses in cases in which unanticipated impacts occur and thus adaptive learning.[76]

Potential Objections to the Legacy Act Concept

There are a number of likely objections to the concept of the Legacy Act and the contours proposed here. One objection to the scope of the Act as proposed herein is that the statute seeks to preserve and constrains solely the use of public natural resources and solely those within the United States. Yet, as a nation, we depend on private as well as public natural resources and on resources found outside as well as within the United States. Certainly, many important resources in the United States are in private hands, and the Legacy Act does not purport to reach or protect these. And many resources on which we depend are imported from around the globe. The criticism of the Legacy Act is that it does not protect all or perhaps even most of the resources on which we currently depend.

Although it is true that many resources we use are outside the direct scope of the Legacy Act, this fact does not negate the value of such a statute. First, the existence of private resources and resources outside the United States that supplement public resources in the United States does not mean that we should fail to take particular care of the natural resources in federal public ownership. These resources have been reserved or acquired by the government to serve the national interest, and in some cases they represent irreplaceable values. They may have both economic value and value that is more difficult to quantify but of great importance to our national identity.[77] For example, the history of the national parks reveals that they were created in part to inspire and serve as symbols for the nation.[78] The history of wilderness areas and national wildlife refuges shows similar awareness of a broad array of noneconomic values.[79] The resources on these lands are uniquely affected with a public interest, and therefore we owe a duty to set policy for them with the broad public interest in view.

Moreover, notwithstanding the fact that we draw resources from sources other than federal public resources, the resources in public ownership in the United States are uniquely in our control and therefore particularly valuable. Taking special steps to monitor our use and degradation of resources within public ownership and control, and to decide consciously what stock of public natural resources we wish to preserve, is a sound policy notwithstanding global trade in some of these resources. Further, the Legacy Act recognizes and takes into account other stocks of resources. The statute provides an exception to the prohibition on impermissible resource depletion or degradation if it can be shown that substitute resources will be available at the conclusion of the legacy period. Thus, for resources with purely economic value, if U.S. supplies represent a small fraction of available resources, and other resources will be available to meet U.S. needs in the future, the duty to preserve the resource is abated.

Finally, as to the exclusion of privately owned resources in the United States, the decision to exclude these from the statute's reach reflects a judgment that the American public has a unique concern for the resources in public ownership and would support greater accountability in the disposition of these resources because they are held for the public good. The idea of a legacy act that applies to natural resources in private ownership would require a considerably different approach. To apply the Act as proposed to all natural resources in private hands would represent a dramatic shift in policy. Even the scale of information collection under such a statute would raise serious questions of cost, feasibility, and not least, intrusion into individual privacy. However, determining the appropriate values to embody in the statute would be an important part of the debate that should precede enactment of such a law.

Another objection sweeps more broadly. One might reject the Legacy Act concept because one rejects the value of retaining public resources. For example, some have argued that public natural resources should be privatized and that private ownership and markets will eliminate all problems associated with allocation of these resources.[80] But it is clear that free markets are not the answer. The combination of externalities associated with natural resource depletion and degradation for which markets fail to account, the lack of markets for many public natural resources and the services and values they provide, and the imperfect information about how proposed actions will affect those resources make it virtually certain that unregulated use of public natural resources will not achieve even the minimal goal that market proponents seek – economic efficiency – much less other societal goals.[81]

Moreover, just as individuals face pressures in favor of short-term consumption of resources, managers of business enterprises that use and degrade resources are subject to pressures to maximize profits in the short term.[82] Our system of corporate governance and finance demands that managers of publicly held companies act as stewards not for the public but for the companies' investors. The quarterly earnings-reporting cycles of capital markets impose an extremely short-term horizon within which managers must demonstrate profitability, and they create enormous pressure for short-term profits.[83] Even a company committed in principle to stewardship of public resources cannot be expected to pursue that value when it is at odds with the short-term economic interests of investors, if the company operates within the discipline of the market. Thus, to achieve protection of a legacy requires that we employ some means to offset the incentives of the market.

The stunning examples provided by corporate leaders such as Ray Anderson, founder and chair of Interface Inc., and Paul Hawken, co-founder of Smith and Hawken and founder and executive director of the Natural Capital Institute, provide ample evidence that, with motivation, today's corporate practices can be radically changed in ways that reduce resource depletion and degradation while improving efficiency and generating private profits.[84] However, the savings are not obvious and require that our ingenuity and creativity be engaged. Because of uncertainty, lack of information, and timidity, markets have failed to capture these benefits, and regulators have hesitated to require practices that would produce them. The Legacy Act would help to stimulate creativity and harness ingenuity in service of our shared goals.

Finally, some may seek to justify the status quo of public natural resource depletion and degradation by declaring their faith in the ingenuity of future generations to develop alternatives to the resources we deplete today. Proponents of the status quo sometimes imply that those who favor preservation of natural resources lack faith in human creativity and ingenuity. This is incorrect. As envisioned, the Legacy Act counts on the ingenuity and creativity of today's generations to find better, more-resource-conserving ways to operate and live rather than assuming that unidentified future generations will have ingenuity we lack today.[85] Moreover, there are many natural resources and values and services associated with them for which there are no substitutes. Once these are depleted, they are gone forever. The Legacy Act rejects as contrary to fundamental American values a policy that assumes we can destroy everything and either replace or learn to do without it.[86]

Conclusion

This chapter sets out a proposal for a new statutory approach that could help us to protect the interests of the present and future generations in our public natural resources. The Legacy Act would replace the open-ended balancing tests that govern many decisions affecting public natural resources under current law with an enforceable mandate. It would also help us to transcend the problems created by the short time horizon that governs decisions of many economic enterprises and by the widely acknowledged flaws of the cost-benefit analysis that drives much governmental decision making.[87] The statute's design would counterbalance the systematic bias in favor of current economic uses of public natural resources and would curtail the attendant depletion and degradation. In the remaining chapters of Part 1, the authors consider some of the questions raised by this proposal, identify additional issues for consideration, and highlight the opportunities the statute might provide.

NOTES

1. *See* Alyson C. Flournoy, Heather Halter & Christina Storz, *Harnessing the Power of Information to Protect our Public Natural Resource Legacy*, 86 Tex. L. Rev. 1575, 1575–77 (2008).
2. *See* Ryan Feinberg, Alyson C. Flournoy, Margaret Clune Giblin, Heather Halter & Christina Storz, The Future of Environmental Protection: The Case for a National Environmental Legacy Act: A Center for Progressive Reform Report 3–4, *available at* http://www.progressivereform.org/articles.

3. *See*, *e.g.*, NOAA's NATIONAL MARINE FISHERIES SERVICE REPORT ON THE STATUS OF THE U.S. FISHERIES FOR 2006, *available at* http://www.nmfs.noaa.gov/sfa/domes_fish/StatusoFisheries/2006/2006RTCFinal_Report.pdf; MILLENNIUM ECOSYSTEM ASSESSMENT, ECOSYSTEMS AND HUMAN WELL-BEING: BIODIVERSITY SYNTHESIS 2–5 (2005), *available at* http://www.millenniumassessment.org/documents/document.354.aspx.pdf; MILLENNIUM ECOSYSTEM ASSESSMENT, ECOSYSTEMS AND HUMAN WELL-BEING: WETLANDS AND WATER SYNTHESIS 2–10 (2005), *available at* http://www.millenniumassessment.org/documents/document.358.aspx.pdf; ENERGY INFO. ADMIN., U.S. DOE, ENERGY IN THE UNITED STATES: 1635–2000, ENERGY OUTLOOK AS OF 2001, *available at* http://www.eia.doe.gov/emeu/aer/eh/frame.html; WELFARE RANCHING: THE SUBSIDIZED DESTRUCTION OF THE AMERICAN WEST, 162–257 (George Wuerthner and Mollie Matteson eds., 2002).
4. *See* ALYSON C. FLOURNOY, MARGARET GIBLIN & MATT SCHUDTZ, SQUANDERING PUBLIC RESOURCES: A CENTER FOR PROGRESSIVE REFORM REPORT (Sept. 2007), *available at* http://www.progressivereform.org/articles/Squndering_Public_Resources.pdf); *see also* Alyson C. Flournoy, *Protecting a Natural Resource Legacy While Promoting Resilience: Can It Be Done*? 87 NEB. L. REV. 1008, 1011–16 (2009).
5. *See* Chapter 2 of this volume.
6. *See* S. Utah Wilderness Alliance v. Norton, 542 U.S. 55 (2004) (precluding suits to compel agency action under § 706(1) of the Administrative Procedure Act if the action is not sufficiently discrete and finding the terms of the resource management plan at issue were merely statements of priority that guide and constrain agency action but do not prescribe or require action); Ohio Forestry Ass'n v. Sierra Club, 523 U.S. 726 (1998) (holding forest management plan not ripe for review).
7. This chapter proposes a federal statute, but the concept and design of the Legacy Act could easily be adapted for adoption as a state statute as well. As with NEPA, state analogues could serve distinct purposes. A state legacy act would presumably focus on protecting a legacy of state-owned and public trust natural resources rather than federal resources.
8. Paul Hawken, Amory Lovins, and L. Hunter Lovins coined the phrase "natural capital" to capture this idea in their groundbreaking book, NATURAL CAPITALISM: CREATING THE NEXT INDUSTRIAL REVOLUTION (1999). *See id.* at 2–6, 148–69.
9. *See*, *e.g.*, Terri Cullin, *Fiscally Fit: Making Sense of Retirement Plans*, WALL ST. J., Sept. 13, 2007, *available at* http://online.wsj.com/public/article/SB118952742492823919.html; Financial Planning Association, *20 Keys to Being a Smarter Investor* 3, *available at* http://www.fpanet.org.
10. As noted previously, some statutes have general mandates for sustainability, but these are undermined by specific authority to consider and balance numerous competing values. The Endangered Species Act is the resource conservation statute with the clearest enforceable standard – the antiextinction mandate – and this structure has proved relatively effective. However, one of its frequently cited shortcomings is that its protections are triggered only when a species is on the brink of extinction.

11. For a discussion of various strong and weak formulations of sustainability, *see* Bryan G. Norton, Sustainability: A Philosophy of Adaptive Ecosystem Management, ch. 8 (2005).
12. *See* Bjørn Lomborg, The Skeptical Environmentalist: Measuring the Real State of the World 159 (2003) (claiming that future generations will be able to allocate more resources toward conservation, thereby raising living standards); Office of Management & Budget, Informing Regulatory Decisions: 2003 Report to Congress on the Costs and Benefits of Federal Regulations and Unfunded Mandates on State, Local, and Tribal Entities (2003) (assuming that future generations will be wealthier and therefore will value marginal benefits and costs less than people alive today).
13. 16 U.S.C. § 529 (2006).
14. 16 U.S.C. § 1851 (2006).
15. 42 U.S.C. § 4321 (2006).
16. Resilience is a way to describe "the persistence of relationships within a system and . . . a measure of the ability of these systems to absorb changes" and still persist. C.S. Holling, *Resilience and Stability of Ecological Systems*, 4 Ann. Rev. Ecology & Systematics 1 (1973). It describes the degree of disturbance a system can tolerate before it flips into another behavior regime. Lance Gunderson, *Resilience, Flexibility and Adaptive Management: Antidotes for Spurious Certitude?* 3 Conservation Ecology 7 (Winter 1999), *available at* http://consecol.org/vol3/iss1/art7. This is discussed further *infra* Ecosystem Resilience: The Bottom Line.
17. The difficulties of identifying the causes of a particular change after the fact in some circumstances would be considerable. However, the Act is primarily a preemptive statute, designed to prevent agencies from permitting future actions that will cause excessive degradation or depletion. Thus, agencies' analysis will typically be prospective – predicting the degree of depletion or degradation that particular proposed actions will cause – rather than seeking retrospectively to identify the cause of degradation or depletion. This is considered further *infra*, The Contours of the National Environmental Legacy Act.
18. The generic use of the term "resources" does not resolve some important questions that would arise in drafting and implementing the Legacy Act. For example, the sketch of the Act herein prohibits impermissible depletion or degradation of covered resources and development of metrics for covered resources. Both usages beg the question of whether, for living resources, the prohibition and the metrics should be interpreted to apply at the organism, species, or ecosystem level or at some combination of these. This question raises both technical and policy issues. As a matter of policy, the goal in selecting among these would be to ensure enforceability of the Act and protection of the mandated legacy.
19. Design of the statute will require both considerable technical work and further elaboration of value choices. This sketch of the statute's contours includes section numbers for ease of reference. However, it is intended as a sketch of the contours of the Legacy Act, not a detailed statutory proposal.

20. A period of thirty to fifty years would seem appropriate to ensure a long-term perspective.
21. The term "person" should be very broadly defined to include all public and private actors. *See, e.g.*, 16 U.S.C. § 1532(13). The statute should make clear that the prohibition on actions that impermissibly degrade or deplete legacy resources applies both to private actors and to agencies whose actions affect the relevant resource – including managing, permitting, and leasing that affects the resource.
22. Where an agency has stewardship responsibilities for a particular resource under existing law, it would seem most efficient to designate that agency for this role, unless experience suggests this would be inconsistent with achieving the purposes of the Act.
23. *See supra* note 16.
24. For those agencies that already undertake planning regarding the relevant resource, this duty should be coordinated with the agencies' planning duties under existing enabling acts to the maximum extent practicable.
25. *See* Chapter 4 in this volume.
26. *See* Chapter 2 in this volume.
27. *See* Chapter 4 in this volume.
28. Flournoy et al., *supra* note 1, provide a more extensive analysis of how the Legacy Act differs from NEPA, focusing specifically on the use of information.
29. James L. Connaughton, *Keynote Address: Modernizing the National Environmental Policy Act: Back to the Future*, 12 N.Y.U. Envtl. L.J. 1 (2003); Council on Envtl. Quality, The National Environmental Policy Act: A Study of Its Effectiveness after Twenty-Five Years (1997), *available at* http://ceq.eh.doe.gov/nepa/nepa25fn.pdf [hereinafter NEPA at Twenty-Five].
30. *See* Dinah Bear, *Some Modest Suggestions for Improving Implementation of the National Environmental Policy Act*, 43 Nat. Resources J. 931 (2003); Lynton K. Caldwell, *Beyond NEPA: Future Significance of the National Environmental Policy Act*, 22 Harv. Envtl. L. Rev. 203 (1998); Connaughton, *supra* note 29; NEPA at Twenty-Five, *supra* note 29; Paul J. Culhane, *NEPA's Impacts on Federal Agencies, Anticipated and Unanticipated*, 20 Envtl. L. 681 (1990); Mary H. O'Brien, *NEPA as It Was Meant to Be;* NCAP v. Block, *Herbicides, and Region 6 Forest Service*, 20 Envtl. L. 735 (1990); Stark Ackerman, *Observations on the Transformation of the Forest Service: The Effects of the National Environmental Policy Act on U.S. Forest Service Decision Making*, 20 Envtl. L. 703 (1990); Bradley C. Karkkainen, *Toward a Smarter NEPA: Monitoring and Managing Government's Environmental Performance*, 102 Colum. L. Rev. 903, 906 (2002).
31. Matthew Bender & Co., 4–9 Treatise on Environmental Law § 9.08 (2007) (describing NEPA's spillover effect, causing state governments to consider the environmental impacts of their actions through "little NEPAs"); Oliver A. Houck, *The Secret Opinions of the United States Supreme Court on Leading Cases in Environmental Law, Never Before Published!* 65 U. Colo. L. Rev. 459, 469–70 (1994) (stating that most states, the European Community, and almost every developed

nation in the world has adopted law similar to NEPA); Bear, *supra* note 30, at 931 (noting that NEPA's "environmental impact assessment process has been adopted by over a hundred countries").

32. 42 U.S.C. § 4331(a).
33. *Id.*
34. *Id.*
35. 42 U.S.C. § 4331(b).
36. 42 U.S.C. § 4331(b)(3).
37. *See, e.g.*, Lynton Keith Caldwell, The National Environmental Policy Act: An Agenda for the Future (1998); NEPA at Twenty-Five; Oliver Houck, *Is That All? A Review of the National Environmental Policy Act: An Agenda for the Future*, 11 Duke Envtl. L & Policy F. 173, 178–79 (2000); Nicholas Yost, *NEPA's Promise – Partially Fulfilled*, 20 Envtl. L. 533 (1990).
38. Strycker's Bay Neighborhood Council, Inc. v. Karlen, 444 U.S. 223 (1980) (holding that NEPA merely requires that agencies consider the environmental consequences of their proposed actions); *see also* Caldwell, *supra* note 37 (NEPA was not intended to be regulatory); Houck, *supra* note 37 (noting the absence of a precise, enforceable policy). This is discussed in greater depth herein.
39. *See* Karkkainen, *supra* note 30, at 903 (2002) (arguing that allowing adaptive management techniques would streamline the NEPA process).
40. Oliver Houck, *Unfinished Stories*, 73 U. Colo. L. Rev. 867 (2002) (naming Scenic Hudson Pres. Conference v. Fed. Power Comm'n, Sierra Club v. Morton, and other cases where heightened public awareness has affected agency decisions); *see* Daniel R. Mandelker, NEPA Law and Litigation 2d § 11:3 (2007) (stating that the threat of interminable delays and litigation under NEPA has caused the cancelation of many projects); William Buzbee, *The Regulatory Fragmentation Continuum, Westway and the Challenges of Regional Growth,* 21 J.L. & Pol. 323, 340–41 (2005) (arguing that, in cases of major land-use actions, federal, state, and local levels of government all become involved, and cases often have underlying motivations).
41. 42 U.S.C. § 4332(C).
42. For a detailed discussion of how NEPA's shortcomings have interfered with its legacy ambitions, *see* Flournoy et al., *supra* note 1.
43. Matthew J. Lindstrom, *Procedures without Purpose: The Withering Away of the National Environmental Policy Act's Substantive Law*, 20 J. Land Resources & Envtl. L. 245, 264 (2000) (claiming that courts have been ambivalent to the practical significance of NEPA's substantive provisions); Oliver A. Houck, *supra* note 37 (stating that NEPA lacks teeth); Bear, *supra* note 30, at 931 (2003) (claiming that, oftentimes, NEPA's requirement to analyze alternatives is counterproductive); Harvey Bartlett, *Is NEPA Substantive Review Extinct, or Merely Hibernating?* 13 Tul. Envtl. L.J. 411 (2000) (arguing that while NEPA's substantive force is not in use, it has the potential to be revived).

44. The documentation takes the form of an environmental assessment, environmental impact assessment, or finding of no significant impact, depending on the extent of the impacts identified. *See* 40 C.F.R. §§ 1501.3–1501.4.
45. This interpretation of NEPA was crystallized early on by the Supreme Court in *Strycker's Bay* (holding that NEPA merely requires that agencies consider the environmental consequences of their proposed actions). *See also* Robertson v. Methow Valley Citizens Council, 490 U.S. 332, 351 (1989) (stating that "NEPA merely prohibits uninformed – rather than unwise – agency action.").
46. Despite the desirability of clarity in the legislation itself, the limitations of the legislative process make it inevitable that regulations will contain important clarifications. The recommendation for appointment of a committee of scientists, discussed herein, is intended to ensure that the regulations are technically well constructed. However, to the extent possible, the value choices embedded in the standard should be defined as clearly as possible in the statute.
47. As is discussed further herein, different standards for renewable and nonrenewable resources would permit us to be able to use some proportion of our nonrenewable resources, such as fossil fuels, today while recognizing that there is less justification to spend down resources that can renew themselves with proper stewardship.
48. World Comm'n on Env't & Dev., Our Common Future 43 (1987).
49. John C. Dernbach, *Sustainable Development as a Framework for National Governance*, 49 Case W. Res. 1, 31 (1998).
50. The prohibition is tempered by two exceptions described in section 11. These would permit otherwise-impermissible degradation or depletion if clear and convincing proof shows (1) that substitute resources are likely to be available by the end of the legacy period (whether through natural regeneration, substitution, or technological advance) or (2) that there is an overriding public interest that warrants present use of the resource, and mitigation of the resource and its associated values and services is possible and is undertaken.
51. Professors Shapiro and Steinzor propose the use of metrics in a different context with a different objective – as a tool to promote agency accountability. Sid Shapiro & Rena Steinzor, *Capture, Accountability, and Regulatory Metrics*, 86 Tex. L. Rev. 1741, 1769 (2008). It would make sense to try to develop metrics that could be used in more than one context, such as for monitoring both agency accountability and compliance with the Legacy Act.
52. *Id.* at 1770.
53. Professors Shapiro and Steinzor emphasize the importance of independence in the context of developing metrics for agency accountability as well. *Id.* at 1775–76.
54. Gunderson, *supra* note 16, at 7 (recommending a shift in our focus from "variables of the moment" like water levels and population numbers to "more enduring system properties such as resilience, adaptive capacity and renewal capability"; Ahjond S. Garmestani, Craig R. Allen & Heriberto Cabezas, *Panarchy, Adaptive*

Management and Governance: Policy Options for Building Resilience, 87 Neb. L. Rev.1036 (2009).

55. *See* Chapters 5–6 in this volume.
56. Karkkainen, *supra* note 30, at 940; *see* Chapter 6 in this volume.
57. David M. Driesen, *Is Cost-Benefit Analysis Neutral?* 77 U. Colo. L. Rev. 335, 339–42 (2006).
58. Flournoy, *supra* note 4, explores the importance of incorporating resilience into the design of the Legacy Act.
59. Holling, *supra* note 16, at 17.
60. Gunderson, *supra* note 16.
61. *Id.*
62. *Id.* at 7.
63. *See* Chapter 3 in this volume.
64. Holly Doremus, *Precaution, Science and Learning While Doing in Natural Resource Management*, 82 Wash. L. Rev. 547 (2007), and *The Purposes, Effects, and Future of the Endangered Species Act's Best Available Science Mandate*, 34 Envtl. L. 397 (2004); Daniel C. Esty, *Environmental Protection in the Information Age*, 79 N.Y.U. L. Rev. 115, 197 (2004); Howard A. Latin, *The "Significance" of Toxic Health Risks: An Essay on Legal Decisionmaking under Uncertainty*, 10 Ecology L.Q. 339 (1982); John S. Applegate, *The Perils of Unreasonable Risk: Information, Regulatory Policy, and Toxic Substances Control*, 91 Colum. L. Rev. 261 (1991).
65. *See* Applegate, *supra* note 64 (detailing the history of failed regulation under the Toxic Substances Control Act); Latin, *supra* note 64.
66. The Intergovernmental Panel on Climate Change (IPCC) developed its first climate change assessment in 1990, which represented scientific consensus that climate change exists and is a global threat. *See* IPCC, Scientific Assessment of Climate change – Report of Working Group I (1990), *available at* http://www.ipcc.ch/pub/reports.htm; *see* IPCC, Impacts Assessment of Climate Change – Report of Working Group II (1990), *available at* http://www.ipcc.ch/pub/reports.htm; *see* IPCC, The IPCC Response Strategies – Report of Working Group III (1990), *available at* http://www.ipcc.ch/pub/reports.htm. Despite these reports, as late as 2001, President George W. Bush's statements on climate change focused on reducing the uncertainty of the science as opposed to reversing the global warming trend. *See, e.g.*, Office of the Press Secretary, *President's Statement on Climate Change* (July 13, 2001), *available at* http://www.whitehouse.gov/news/releases/2001/07/20010713-2.html (announcing various projects aimed to reduce the uncertainty surrounding the science of climate change). *See* Office of the White House, Global Climate Change Policy Book, Executive Summary (Feb. 2002), *available at* http://www.whitehouse.gov/news/releases/2002/02/climatechange.html (illustrating the shift in climate change discourse of the executive office from reducing uncertainty to taking action). These claims were then amplified by the media's approach of reporting on the existence of an alleged scientific debate on this issue. *See* Jules

Boykoff & Maxwell Boykoff, *Journalistic Balance as Global Warming Bias: Creating Controversy Where Science Finds Consensus*, FAIRNESS & ACCURACY IN REPORTING (Nov./Dec. 2004), *available at* http://www.fair.org/index.php?page=1978 (arguing that, by focusing on the existence of a debate about global warming rather than evaluating the strengths of the science on either side of the debate, the media portrays the existence and effects of climate change to be uncertain).

67. John S. Applegate, *The Taming of the Precautionary Principle*, 27 WM. & MARY ENVTL. L. & POL'Y REV. 13 (2002) (defining precautionary principle to embody principles that anthropogenic harm to human health and the environment be avoided or minimized through anticipatory, preventive regulatory controls; and to accomplish this, activities and technologies whose environmental consequences are uncertain but potentially serious should be restricted until the uncertainty is largely resolved).
68. Cass R. Sunstein, *Beyond the Precautionary Principle*, 151 U. PENN. L. REV. 1003 (2003) (arguing that, in its strongest form, the precautionary principle fails to offer any guidance on the proper course of action).
69. For example, any individual action that would be deemed significant under NEPA might be prohibited, pending collection of the relevant information and development of agency metrics. Defining the precise class of activities to be proscribed in this transition period would entail significant value choices, and resource users would likely strongly oppose such a provision. However, it would be extremely important for Congress to resolve this in favor of a provision that ensures the statute's efficacy. Without such a provision, resource users would have the incentive and opportunity to delay implementation of the Act's protections by challenging the adequacy of agencies' information at every step in the process.
70. Wendy E. Wagner, *Commons Ignorance: The Failure of Environmental Law to Produce Needed Information on Health and the Environment*, 53 DUKE L.J. 1619, 1741–42 (2004).
71. *Id.*
72. Bradley C. Karkkainen, *Logjams and Baselines: Tackling Information Deficits in Environmental Regulation*, 86 TEX. L. REV. 1409, 1427–34 (2009).
73. Exceptions or exemptions should be employed to achieve balance between the goals of the Legacy Act and other public values in tension with its purposes. Obviously, the broader the scope of any exceptions or exemptions, the less effective the Legacy Act would be at achieving its purposes. The danger of permitting overly broad and arguably unnecessary exceptions and exemptions for ostensibly compelling purposes such as national security has been demonstrated by recent experience with NEPA. *See* Testimony of Michael Anderson, Senior Resource Analyst, Wilderness Society, to H. Resources Comm. Nat'l Envtl. Pol'y Act Task Force, Field Hearing, *The Role of NEPA in the States of Washington, Oregon, Idaho, Montana and Alaska*, Spokane, Wash., April 23, 2005, *available at* http://www.wilderness.org/Library/Documents/upload/Testimony-Anderson-NEPA-20050423.pdf (arguing that, through various categorical exclusions, the Bush administration

limited "public participation and environmental consideration in federal land management").

74. Karkkainen, *supra* note 30, at 925–32.
75. *Id.* at 938–39.
76. Holly Doremus, *Precaution, Science and Learning While Doing in Natural Resource Management*, 82 WASH. L. REV. 547, 549–50 (advocating an approach that is precautionary and incorporates adaptive learning); Karkkainen, *supra* note 30, at 938–40 (2002) (arguing that allowing adaptive management techniques would streamline the NEPA process).
77. *See* Chapter 11 in this volume.
78. Holly Doremus, *Nature, Knowledge and Profit: The Yellowstone Bioprospecting Controversy and the Core Purposes of America's National Parks*, 26 ECOLOGY L.Q. 401, 437–43 (1999); *see* Chapter 11 in this volume.
79. *See* Rob Fischman, *The Significance of National Wildlife Refuges in the Development of U.S. Conservation Policy*, 21 J. LAND USE & ENVTL. L. 1 (2005). Fischman explains that a variety of motivations sparked creation of individual refuges, but all were generally created for conservation of animal species. In providing for creation of wilderness areas, Congress explicitly sought to preserve the "wilderness character" of lands. 16 U.S.C. § 1131.
80. *See, e.g.*, Richard L. Stroup, *Privatizing Public Lands: Market Solutions to Economic and Environmental Problems*, 19 PUB. LAND & RESOURCES L. REV. 79 (1998) (advocating privatization to improve environmental quality); James Huffman, *Environmental Protection and the Politics of Property Rights: The Public Interest in Private Property Rights,* 50 OKLA. L. REV. 377, 378 (1997) (arguing that privatization of scarce resources is necessary to protect the resources and people's enjoyment of them).
81. Amy Sinden, *The Tragedy of the Commons and the Myth of a Private Property Solution*, 78 U. COLO. L. REV. 533 (2007).
82. *See* MARK SAGOFF, THE ECONOMY OF THE EARTH 63–65 (1988); JAMES GUSTAVE SPETH, THE BRIDGE AT THE EDGE OF THE WORLD 58–60 (2008).
83. *See* Joshua Frank, *Natural Selection, Rational Economic Behavior, and Alternative Outcomes of the Evolutionary Process*, 32 J. SOCIO-ECONOMICS 601–22 (2003) (arguing that the inefficient practice of pursuing short-term profitability at the expense of long-term growth is prevalent in the United States and other developed nations); James R. Repetti, *Corporate Governance and Stockholder Abdication: Missing Factors in Tax Policy Analysis*, 67 NOTRE DAME L. REV. 971, 980–83 (1992) (arguing that "management has an economic incentive to focus on the short-term" and will do so without stockholder oversight).
84. RAY ANDERSON, MID-COURSE CORRECTION: TOWARD A SUSTAINABLE ENTERPRISE: THE INTERFACE MODEL (1999); PAUL HAWKEN, THE ECOLOGY OF COMMERCE: A DECLARATION OF SUSTAINABILITY (1993); HAWKEN ET AL., *supra* note 8.
85. In fact, the Legacy Act creates an incentive for technological development of alternatives to nonrenewable resources by providing an exception that will allow greater

depletion of a particular resource on proof that an alternative that provides all relevant services and values will be available at the end of the legacy period.

86. *See* Chapter 11 in this volume; Douglas A. Kysar, *Fish Tales*, in RESOURCES FOR THE FUTURE, REFORMING REGULATORY IMPACT ANALYSIS 200–02 (Winston Harrington, Lisa Heinzerling & Richard D. Morganstern eds., 2009) (describing the distinction between the view of resources as substitutable embedded in cost-benefit analysis and the view of resources as irreplaceable embedded in ecological economics).

87. *See generally* David Driesen, *The Economic Dynamics of Environmental Law: Cost-Benefit Analysis, Emissions Trading, and Priority-Setting*, 31 B.C. ENVTL. AFF. L. REV. 501, 504–06 (2004) (arguing that cost-benefit analysis is ineffective because of the difficulty of quantifying the costs of even well-understood dangers); *see generally* Lisa Heinzerling, *Regulatory Costs of Mythic Proportions*, 107 YALE L.J. 1981 (1998) (criticizing use of discounting in cost-benefit analysis of Occupational Safety and Health Administration regulations); FRANK ACKERMAN & LISA HEINZERLING, PRICELESS: ON KNOWING THE PRICE OF EVERYTHING AND THE VALUE OF NOTHING 44–53 (2004) (arguing that cost-benefit studies of regulatory costs are flawed because they (1) consider regulations that were never adopted, (2) ignore risks and benefits other than life and death, and (3) use discounting, which dismisses long-term risks).

2 The Necessity of Procedural Reform

Sidney Shapiro

"IF YOU LET ME WRITE PROCEDURE AND I LET YOU WRITE substance," Congressman John Dingell once observed, "I'll screw you every time."[1] The environmental movement in the 1960s was well aware of the important connection between legal procedure and the successful implementation of the environmental statutes. These activists convinced the courts to adopt a series of reforms that made agencies such as the Environmental Protection Agency (EPA) more accountable for failing to implement a statutory mandate. More recently, the Supreme Court has made it more difficult for environmental advocacy groups to perform this role. If the National Environmental Legacy Act (hereafter, "the Act") is to achieve its objectives, Congress should consider how administrative procedure will affect its substantive objectives. Although substantive reform is important, as Representative Dingell reminds us, so is procedural reform.

Procedure and Substance

Environmental and consumer activists in the 1960s and 1970s were concerned that the basic procedural framework used by agencies was insufficient for ensuring that the many new environmental and consumer laws enacted during the period would be implemented effectively. This concern was based in part on a series of widely publicized reports that identified overly cozy relationships between regulators and regulatees as a primary reason for the faltering performance of older regulatory agencies, such as the Federal Trade Commission (FTC).[2] Reformers attributed this capture in part to the way in which the courts had interpreted the Administrative Procedure Act (APA).[3] Because the courts focused almost exclusively on how the APA protected regulated entities, it furnished only limited procedural rights for regulated beneficiaries.

A sympathetic judiciary, which expanded the procedural rights of regulatory beneficiaries, produced a "reformation" of administrative law.[4] The reformation worked in favor of regulatory beneficiaries in several ways. The courts expanded rule-making notice requirements in a way that gave environmental groups the opportunity to contest the evidence used by agencies to justify inadequate regulation. A presumption in favor of judicial review made it more likely that agencies could be sued when they failed to regulate adequately. The courts liberalized standing requirements, thereby expanding the eligibility of regulatory beneficiaries to sue the EPA and other agencies for the failure to implement environmental policies effectively. The courts also authorized environmental groups to sue on behalf of their members – thus creating a pool of plaintiffs with the resources to sue and the negotiating leverage to bargain with agencies for stronger environmental protections.

More recently, the courts have become more hostile to regulatory beneficiaries, a change I have described as the "counter-reformation" of administrative procedure.[5] Among other actions, the Supreme Court has ruled that some challenges to inadequate agency action are not ripe for review, and it has found that regulatory beneficiaries lack standing to challenge agency decisions. In addition, judges have been reluctant to second-guess an agency when it refuses to act.

If Congress is to ensure that the Act achieves its intended purposes, it has to address each of these three problems. In the Act, Congress should make natural-resource-planning documents subject to immediate review, establish deadlines for agency action under the Act, identify who is entitled to bring a lawsuit to enforce the provisions of the act, and articulate how those persons are harmed by the lack of effective regulatory action. Congress should also award a cash prize to plaintiffs who establish that some agency action or inaction is illegal under the Act. The remainder of this chapter explains the necessity of these legislative provisions and assesses whether they will overcome the roadblocks established by the Supreme Court that prevent regulatory beneficiaries from holding agencies accountable.

Ripeness and Planning Documents

The Act will require appropriate agencies to develop plans for ensuring the preservation of natural resources for generations to come. If an environmental interest group believes that a government agency has adopted a plan that fails in some manner to live up to the statutory mandate, the most effective recourse would be to attempt to sue the agency and argue to the court that the plan is illegal. Current Supreme Court decisions, if left in place by Congress, may prevent such review.

In *Lujan v. National Wildlife Federation*,[6] environmental groups sought to challenge a plan adopted by the Department of the Interior to guide a review of what types of activities would be permitted on 180 million acres of public lands. The intent of the plan was to increase resource-extraction activities, such as logging, where those activities were currently prohibited. Whether logging or other activities would be permitted was to be determined by an application of the criteria in the plan. The National Wildlife Federation sought to challenge the plan on the ground that it was inconsistent with the Federal Land Policy and Management Act (FLPMA), which was the source of the agency's legal authority for adopting the plan. The Supreme Court, however, found that the suit was not ripe for review.

The Supreme Court created the current ripeness doctrine in *Abbott Laboratories v. Gardner*,[7] which involved a company's attempt to obtain judicial review of a Food and Drug Administration (FDA) regulation before the FDA had enforced the regulation. Before *Abbott Laboratories*, a regulated entity could challenge the legality of a new regulation only in the context of an attempt by an agency to enforce the rule. The rationale for this restriction was to "prevent the courts, through avoidance of premature adjudication, from entangling themselves in abstract disagreements over administrative policies, and also to protect the agencies from judicial interference until an administrative decision has been formalized and its effects felt in a concrete way by the challenging parties."[8] In *Abbott Laboratories*, the Court recognized that withholding review until a rule is enforced could work a hardship on regulated entities because a company has to decide whether to obey a regulation. If it obeyed the regulation and it turned out later that the rule was invalid, the company would have wasted resources in complying with an invalid rule. But if the company decided not to obey the regulation, it could end up paying fines for that failure if the rule was upheld. The Court adopted a new approach to ripeness that addressed this Hobson's choice. Ripeness is now a two-pronged approach that balances the hardship to the litigant of withholding review with the benefit to the court of waiting until an enforcement action to hear arguments about the illegality of the rule.

In *Lujan v. National Wildlife Federation*, the Court found that the department's plan was not ripe for review before it was actually used to make specific resources decisions because of the benefits of waiting for review noted by *Abbott Laboratories*:

> Some statutes permit broad regulations to serve as the "agency action," and thus to be the object of judicial review directly, even before the concrete effects normally required for APA review are felt. Absent such a provision, however, a regulation is not ordinarily considered the type of agency action "ripe" for judicial review under the APA until the scope of the controversy has been reduced to more manageable proportions, and its factual components fleshed out, by some concrete action applying the regulation to the claimant's situation in a fashion that harms or threatens to harm him.[9]

In *Ohio Forestry Association, Inc. v. Sierra Club*,[10] the Court confirmed that resource-planning documents were not ripe for review unless Congress had specified otherwise. In the case, the plaintiff sought to challenge a plan that specified criteria to be used to make decisions about timber production in a national forest. According to the Court, "Although the [forest management] Plan sets logging goals, selects the areas of the forest that are suited to timber production . . . and determines which 'probable methods of timber harvest' are appropriate, . . . it does not itself authorize the cutting of any trees."[11] The Court therefore ruled that the plaintiff could not challenge the plan until the agency had issued a permit to harvest trees in specific areas of the forest.

Environmental plaintiffs apparently have not prevailed because they cannot claim the same type of hardship suffered by regulated entities. *Abbott Laboratories* focused on the Hobson's choice of regulated entities on whether to comply with a possibly illegal regulation. No environmental plaintiffs face this type of choice because they are not required to obey the plan or risk fines if they do not. This does not mean, however, that refusing immediate judicial review of planning documents poses no hardship for environmental plaintiffs. It is more difficult and expensive for them to mount legal challenges to specific decisions made pursuant to a plan than it is to attack its legality in one action. No court, however, has considered this hardship sufficient to outweigh the advantage of postponing judicial review until the plan is used in specific resource decisions. Moreover, the Supreme Court has suggested that to qualify as a "hardship," the hardship must involve "'adverse effects of a strictly legal kind.'"[12]

The difficulty with the Court's approach to ripeness is that it fails to consider the benefit to the public of obtaining an earlier and complete determination on the question of whether a planning document is inconsistent with the law under which it was adopted or with other applicable laws. As compared to piecemeal litigation over different parts of a plan when used to make specific resource decisions, preimplementation review offers the opportunity to vet an entire plan at once. Therefore, such review is more likely to ensure overall compliance with the relevant statute. Particularly in the context of the Act, which challenges agencies to ensure that cumulative impacts of activities do not exceed specified levels of resource degradation or depletion, this advance review would be essential. The availability of immediate review also makes it less likely that the policies adopted in a plan become entrenched as a result of the agency's reliance on them in the period before a lawsuit over a specific application of the plan winds its way through the courts – a process that may take several years after the plan's adoption. A requirement of immediate review also recognizes that, because environmental groups have limited resources, they will be in a better position to challenge illegality if they can do so in one lawsuit instead of multiple lawsuits. Finally, agencies may be less likely to adopt legally dubious plans if they know that such provisions are subject to immediate review instead of piecemeal review at some future unspecified time.

Although there are important benefits to postponing judicial review, they do not outweigh the benefits of immediate review. Congress should therefore include in the Act a provision permitting immediate review of a planning document when it is completed or when it is applied in a specific decision-making context.

Lack of Action and Judicial Review

The fate of the Act depends in no small part on how energetically it is implemented by the administration in power. There is considerable evidence that agencies have been slow and reluctant to fulfill their statutory responsibilities when the administration in office is hostile to regulatory government. One only has to think of the numerous antiregulatory positions of the George W. Bush administration, including, for example, its unwillingness to use existing statutory authority to address global warming. In addition, there is a substantial literature that predicts that agencies will be captured by business interests, given the strong incentives of regulated entities to organize politically to stymie regulatory initiatives they disfavor (or to use government for their benefit) and the weak incentives of individuals to organize to oppose such industry efforts.[13] This claim is overstated; there is considerable evidence that agencies often operate in the public interest.[14] Nevertheless, it is important that environmental plaintiffs can seek judicial review of agency inaction.

The courts review agencies' failure to act according to a process established by the APA. For cases in which the desired action is issuance of a rule, the APA authorizes any interested person to file a petition asking an agency to commence rule making.[15] Once the petition is filed, the APA mandates that the agency respond to the petition within a reasonable time.[16] If the agency fails to respond to the petition, the courts are authorized to "compel agency action... unreasonably delayed."[17] Judges, however, normally defer to an agency's agenda setting and refuse to order the agency to respond to the petition. Courts are somewhat less deferential if a significant number of years have passed since the petition was filed, but they normally do not force a response before the expiration of at least several years. Even in cases of excessive delay, a court will typically ask the agency for a timetable on when it can respond, thereby adding additional delay, rather than ordering an agency to respond immediately.[18] Environmental advocates face a distinct but similarly uphill battle when they seek to challenge the denial of a petition in court. When an agency denies a petition, the courts use a highly deferential standard of review.[19]

The failure of the courts to police the lack of agency action reflects two realities. First, in these contexts, judges lack a rule-making record that could be used to scrutinize the agency's failure to respond to a petition or its rejection of a petition. Because the agency has refused to engage in rulemaking, it has not complied with procedures required by the APA to justify a proposed rule, to take public comments on that justification, or to respond to comments when the final rule is issued. In such a situation, it is difficult for the proponent of the petition to persuade a court that it should reject the agency's position.[20] Second, the highly deferential review that is applied reflects court's reluctance to interfere with what are regarded as agency prerogatives concerning priority setting.[21]

This is a valid concern, but the result is that regulatory beneficiaries lack an effective remedy when an agency fails to implement its statutory mandate. There is a real danger that this fate will befall the Act, a statute that will be controversial with conservative administrations and their business allies. Political oversight can sometimes force agencies to do their jobs, but it is an imperfect remedy. The combination of a Republican president and Congress during the George W. Bush administration, for example, resulted in no meaningful oversight of agency inaction.

When Congress faced a similar problem with the EPA in the 1980s, it established statutory deadlines by which time the EPA had to complete certain types of administrative action. Examples of such deadlines can be found in the 1984 amendments to the Resource Conservation and Recovery Act (RCRA);[22] the 1986 amendments to the Comprehensive Environmental Response, Compensation, and Liability Act (CERCLA),[23] the Safe Drinking Water Act (SDWA),[24] and the Toxic Substances Control Act (TSCA);[25] the 1987 amendments to the Clean Water Act (CWA);[26] and the 1988 amendments to the Federal Insecticide, Fungicide, and Rodenticide Act (FIFRA).[27]

Deadlines would have a number of advantages in terms of promoting timely implementation of the Act. Deadlines can assist an agency in blocking political and interest-group pressure to avoid reaching a decision. They can also make the agency more comfortable in resolving difficult issues than it would be in the absence of an action-forcing statutory timetable. A deadline provides the agency – and the courts – an indication of congressional priorities, and thus helps to guide decisions about the allocation of limited agency resources. Deadlines can also encourage more effective legislative oversight because they make it easier for Congress to determine whether an agency has failed to act expeditiously. At the same time, Congress is more likely to pressure an agency to complete a matter if there is a deadline to do so, as compared to an open-ended statutory obligation, because a deadline constitutes a public commitment to achieve a certain end.

Statutory deadlines also put environmental advocates in a better position when they sue an agency for inaction. The existence of a deadline does not necessarily mean that the courts will force an agency to act, but the case law indicates that the existence of statutory deadlines makes it more likely that a court will find an agency's delay unreasonable and will force the agency to remedy that delay. The reason is that the deadline puts the judge in a different position than in cases where no such deadline exists. As discussed earlier, judges are reluctant to order agencies to respond to a petition because they do not want to involve themselves in agency priority setting. Where, by comparison, Congress has established priorities by inserting deadlines into legislation, the judge is asked to enforce a priority-setting decision established by Congress, a request that the judge should honor.

Standing and Judicial Review

Congress can authorize immediate judicial review of planning documents, and it can set deadlines to ensure that agencies undertake the activities required by the Act in a timely manner. These actions may not be enough, however, to ensure that environmental plaintiffs can sue an agency. A litigant can invoke a congressional authorization to sue only if he or she has standing to challenge that action. To have standing, the plaintiff must establish, among other requirements, that the challenged action has caused an injury of some type, which need not be economic, to him or her. This requirement is based on article III of the Constitution, which limits the power of federal courts to the resolution of "cases" and "controversies," a requirement that exists to prevent the courts from giving advisory opinions.

To set limits on which kinds of injury suffice for standing and which do not, the Court uses a set of adjectives (an injury must be both "concrete and particularized" and "actual or imminent") among other requirements.[28] Deploying these requirements, the Court has decided that environmental plaintiffs lacked standing in a number of prominent cases, including *Lujan v. National Wildlife Federation*,[29] discussed earlier. The plaintiffs in that case asserted that they were injured by the agency's program of reclassification because the plan would decrease their recreational and aesthetic enjoyment of the public lands to which the plan applied. Although the Court acknowledged that such injuries could support standing, it also ruled that the plaintiffs did not qualify for standing because they failed to prove that they used the specific lands that the agency might reclassify for such purposes and therefore did not suffer injury.

The difficulty of obtaining standing is further indicated by *Lujan v. Defenders of Wildlife*,[30] where the plaintiffs sought judicial review of a rule exempting projects funded by federal agencies overseas from review under the Endangered Species Act (ESA). That Act requires the federal agencies to follow statutorily described procedures when their actions may jeopardize the continued existence of an endangered species or adversely modify its habitat. The plaintiffs alleged that the funding agency's failure to follow the requisite procedures made it more likely that endangered animals would become extinct. They asserted the extinction of the animals caused them three types of broad injuries. Under an ecosystem-nexus injury argument, they argued that any person who uses any part of a contiguous ecosystem is injured by the extinction of a species even if the use is located some distance away. Under an animal-nexus injury argument, they argued that the extinction of a species injures anyone who derives pleasure from observing members of that species anywhere, as, for example, in a zoo. Finally, the plaintiffs argued that there was a professional-nexus injury because anyone who makes his or her living by studying a species is injured by its extinction. Six members of the Court refused to grant standing on any of these claims, and four of these rejected all three nexus theories as "beyond all reason" and as "pure speculation and fantasy."[31]

The plaintiffs also sought standing by arguing that they would not be able to travel abroad to view the species if it became extinct. Although the same six members of the Court acknowledged that this form of injury might qualify someone for standing, the allegation failed in this case because the injury was not "imminent."[32] The injury was not imminent because the plaintiffs had failed to prove that they had planned to observe the species in its habitat at a date certain in the near future.

Under *Lujan*, plaintiffs must be able to prove that the government's failure to obey the Act has caused a concrete and immediate injury to the litigant, such as by reducing the litigant's recreational or aesthetic use of a particular area of public lands. The government's actions under the Act, however, may not have any significant immediate impact if the purpose and effect of a regulation is to prevent the gradual dilution of public resources over many decades. As a result, any current plaintiff may have difficulty convincing a court that it has suffered a concrete and immediate injury.

A later case, *Massachusetts v. EPA*,[33] suggests that Congress is able to influence how the courts approach standing. In this case, the State of Massachusetts and other plaintiffs appealed a decision by the EPA not to initiate a rule making to regulate emissions of global warming gases (carbon dioxide) by automobiles. Five members of the Court held that Massachusetts had standing because it was injured through the erosion of its coastline caused by global warming, even though the immediate injury suffered by the state was relatively minor. Four dissenting justices concluded that the injuries alleged by the state were insufficient to gain standing because, among other reasons, they were not "actual and imminent," "real and immediate," or "certainly impending."[34]

In defending its decision, the majority noted that Congress had authorized such challenges to EPA actions, a fact it found "to be of critical importance to the standing inquiry."[35] The opinion explained: "'Congress has the power to define injuries and articulate chains of causation that will give rise to a case or controversy where none existed before."[36] When Congress exercises this power, however, it "'must at the very least identify the injury it seeks to vindicate and relate the injury to the class of persons entitled to bring suit.'"[37]

Congress should therefore identify in the Act who is entitled to bring a lawsuit to enforce its provisions, and it should articulate why such persons are harmed by the failure of the government to obey the Act. If Congress establishes a plausible chain of causation, *Massachusetts v. EPA* suggests that the Court will defer to its judgment that the plaintiffs have suffered injuries that will also occur in the future.[38]

Whether this approach will be successful hinges on two things. First, *Massachusetts v. EPA* warns the courts will not "'entertain citizen suits to vindicate the public's nonconcrete interest in the proper administration of the laws.'"[39] Congress must therefore be able to articulate a plausible injury to current plaintiffs. Second, the Court justified its decision in part on the ground that the courts should treat standing claims by states with particular deference.[40] It is not clear to what extent this deference justified the Court's decision to grant standing to Massachusetts despite its small injury and tenuous claim of causation or whether the Court would grant similar deference to individual plaintiffs.

Congress may have another option for ensuring that plaintiffs have standing to sue to enforce the Act. Commentators, led by Cass Sunstein, have suggested that Congress can obtain standing for plaintiffs by awarding a bounty – $500 in Sunstein's proposal – to be paid to environmental litigants who are successful in their lawsuits.[41] The idea that payment of a bounty would give standing to environmental plaintiffs is based on the long history of qui tam lawsuits. In qui tam statutes, Congress authorizes a financial award to any citizen who successfully sues a private party who has violated federal laws in a manner that does financial harm to the government. The False Claims Act, for example, authorizes any citizen who successfully sues a contractor that has defrauded the government to obtain 50 percent of any recovery.[42] The Supreme Court held that a citizen had standing to bring a lawsuit under this legislation.[43]

The potential of earning a financial recovery gives an environmental plaintiff a concrete interest in winning the lawsuit, which would appear to satisfy the requirement in article III that the plaintiff have more than an abstract interest in the outcome of a case. Nevertheless, the Court has noted that "the same might be said of someone who has placed a wager upon the outcome."[44] What is required, according to the Court, is "obtaining compensation for, or preventing, the violation of a legally protected right."[45] This distinction explains why a qui tam plaintiff suing under the FCA has standing. The FCA is a partial assignment of the government's damages claim. In other words, the government itself had a legal right to be compensated when fraud occurred, and the FCA assigned part of that right to the qui tam plaintiff. An environmental plaintiff seeking a $500 bounty would be in a different position than a plaintiff suing under the FCA. The environmental plaintiff would not be asserting a claim against a private individual to recover money that the person might owe the government. Put another way, if Congress created a bounty for winning the lawsuit, it would not be assigning a property interest.

To overcome this problem, Harold Feld recommends that Congress mimic the FCA in other statutes. He proposes a law that would permit any citizen to challenge the legality of proposed or continuing agency action or inaction by filing a petition with an agency.[46] The law would require an agency to respond to the complaint within ninety days or some similar period of time. If the agency concluded that the complaint had merit, it would publicly announce that result, award the petitioner a monetary award, and adjust its behavior to eliminate the illegality asserted in the complaint. If the agency, in contrast, concluded the complaint had no merit, it would issue a statement to that effect. At this point, the petitioner could file a lawsuit challenging the agency's decision, and if it prevailed in the lawsuit, the court would award the petitioner the money denied by the agency. Feld argues that such a law appears to satisfy the requirement that, to have standing, a plaintiff must be "obtaining compensation for . . . the violation of a legally protected right."[47]

The Supreme Court has warned, however, that an "an interest that is merely a 'byproduct' of the suit itself cannot give rise to a cognizable injury in fact for Article III standing purposes."[48] The Court cited the reimbursement of legal fees as an example of such a byproduct interest. If the Court characterized the payment of the money as a reimbursement of petition or litigation expenses, Feld's proposal would not lead to standing for environmental plaintiffs.

Congress lacks any sure way to guarantee that plaintiffs who seek to challenge agency action or inaction under the Act have standing. Congress cannot authorize plaintiffs to sue if they do not present a case and controversy, but it can increase the likelihood that the courts will find standing by articulating in the legislation who can sue and how they are injured by inadequate agency action or agency inaction. Congress can also attempt to facilitate standing by creating a legal right to a monetary award if a plaintiff establishes that an agency that has acted in a manner inconsistent with its legal mandate. To be safe, Congress should adopt both strategies.

Conclusion

Procedure matters. If the Act is to be successfully implemented, the Court must address three aspects of administrative procedure that could frustrate its implementation. Unless Congress requires the courts to hear challenges to resource plans developed under the Act at the time that they are completed, the courts may not permit such challenges. Although environmental advocates will still be able to challenge the legality of such plans at the time they are used to make specific resources decisions, this approach is less likely to ensure that the plans are consistent with the legal mandates of the Act. Unless Congress establishes statutory time deadlines for completion of key aspects of the Act, the courts are unlikely to require agencies to implement those aspects of the Act in a timely manner. Finally, unless Congress takes steps to increase the likelihood that persons who seek to challenge agency action or inaction have standing, the courts may not find there is a case and controversy. Congress should therefore indicate who can sue and how they are injured by inadequate agency action or agency inaction, and it should create a legal right to a monetary award if a plaintiff establishes that an agency that has acted in a manner inconsistent with its legal mandate.

NOTES

1. John McLaughlin, *Detroit Powerhouse – John D. Dingell*, NATIONAL REVIEW, Feb. 27, 1987, *available at* http://findarticles.com/p/articles/mi_m1282/is_v39/ai_4713624.
2. *See, e.g.*, COMM'N TO STUDY THE FTC, AM. BAR ASS'N, *Report of the Commission to Study the Federal Trade Commission* (Sept. 15, 1969); EDWARD FINCH COX JR., ROBERT C. FELLMETH & JOHN E. SCHULTZ, *The Nader Report on the Federal Trade Commission* (1969).
3. 5 U.S.C. §§ 551–59.
4. Richard B. Stewart, *The Reformation of American Administrative Law*, 88 HARV. L. REV. 1667 (1975).
5. *See* Sidney A. Shapiro, *Administrative Law after the Counter-Reformation: Restoring Faith in Pragmatic Government*, 48 U. KAN. L. REV. 689, 694 (2000).
6. Lujan v. Nat'l Wildlife Fed'n, 497 U.S. 871 (1990).
7. Abbott Labs. v. Gardner, 387 U.S. 136 (1967).
8. *Id.* at 148–49.
9. Lujan v. Nat'l Wildlife Fed'n, 497 U.S. at 891.
10. Ohio Forestry Ass'n, Inc. v. Sierra Club, 523 U.S. 726 (1998).
11. *Id.* at 729.
12. National Park Hosp. Ass'n. v. Dept. of Interior, 538 U.S. 803, 809 (2003).
13. *See* MANCUR OLSEN, THE LOGIC OF COLLECTIVE ACTION: PUBLIC GOODS AND THE THEORY OF GROUPS (1969).
14. *See* STEVEN P. CROLEY, REGULATION AND PUBLIC INTERESTS: THE POSSIBILITY OF GOOD REGULATORY GOVERNMENT (2008).
15. 5 U.S.C. §553(e) ("Each agency shall give an interested person the right to petition for the issuance, amendment, or repeal of a rule.").
16. 5 U.S.C. §555(b) (mandating that "within a reasonable time each agency shall conclude a matter").
17. 5 U.S.C. §706(a) ("The reviewing court shall . . . (1) compel agency action unlawfully withheld or unreasonably delayed").
18. *See* Sidney A. Shapiro & Robert L. Glicksman, *Congress, the Supreme Court, and the Quiet Revolution in Administrative Law*, 1988 DUKE L.J. 819, 834.
19. *See, e.g.*, Nat'l Customs Brokers & Forwarders Ass'n of Am., Inc. v. United States, 883 F.2d 93, 96 (D.C. Cir. 1989) (noting that, while failure by an agency to

promulgate a rule is subject to judicial review, such review should be extremely limited and highly deferential); WWHT, Inc. v. FCC, 656 F.2d 807, 818 (D.C. Cir. 1981) (contrasting the broad discretionary power that agencies possess when deciding whether to promulgate rules with the narrow scope of judicial review of such decisions).

20. *See* Richard J. Pierce, Sidney A. Shapiro & Paul R. Verkuil, Administrative Law And Process 218 (4th ed. 2004) ("A court can know only a small fraction of elements that must enter into an agency's process of setting its agenda and allocating its resources among competing tasks.").
21. *See* Norton v. S. Utah Wilderness Alliance, 542 U.S. 55, 66 (2004) (explaining that the principal purpose of the APA's limitations on remedies for agency inaction is to preserve lawful agency discretion from judicial usurpation).
22. *See, e.g.*, 42 U.S.C. § 6924(c)(2) (setting deadline for EPA promulgation of regulations to minimize disposal of containerized liquid hazardous wastes in landfills); § 6924(g) (setting schedule for reviewing listed hazardous wastes and determining methods of land disposal protective of human health and the environment); § 6924(w) (setting deadline for EPA to promulgate final standards for issuing permits for underground storage tanks that cannot be entered for inspection); § 6923(c) (setting deadline for EPA promulgation of standards applicable to transporters of fuel produced from hazardous waste).
23. *See, e.g.*, 42 U.S.C. § 9605(b)-(c) (deadlines for president to revise National Contingency Plan and promulgate amendments to hazard ranking system); § 9604(i)(6)(A) (deadlines for performance of health assessments by administrator of Agency for Toxic Substances and Disease Registry for each facility on National Priorities List (NPL)); § 9604(i)(2)-(3) (deadlines for issuing list of at least one hundred hazardous substances most commonly found at facilities on the NPL, for adding more hazardous substances to that list, and for preparing toxicological profiles of listed substances); § 9602(a) (deadline for EPA regulations establishing reportable quantities for hazardous substances); § 9616(b)–(e) (schedules for evaluating facilities listed in Comprehensive Environmental Response, Compensation, and Liability Information System (CERCLIS), for commencing remedial investigations and feasibility studies for facilities on NPL, and for commencing remedial actions at such facilities); 29 U.S.C. § 655(a) (1982 & Supp. IV 1986) (deadline for secretary of labor to issue standards for health and safety protection of employees engaged in hazardous waste operations).
24. *See, e.g.*, 42 U.S.C. § 300g-1(b)(1) (setting schedule for EPA promulgation of national primary drinking-water regulations for specific contaminants); § 300g-1(b)(7)(C)(i) (setting deadline for EPA promulgation of "criteria under which filtration ... is required as a treatment technique for public water systems supplied by surface water sources"); § 300h-5(a) (setting deadline for EPA modification of groundwater-monitoring regulations for Class I injection wells).
25. *See* 15 U.S.C. § 2643(a) (deadlines for EPA promulgation of regulations for removal of asbestos from schools).

26. *See, e.g.*, 33 U.S.C. § 1314(m) (schedule for EPA publication of plan for annual review and revision of effluent guidelines); § 1342(p)(4) (deadlines for EPA issuance of regulations setting forth permit application requirements for storm-water discharges); § 1342(p)(6) (deadline for EPA issuance of regulations establishing program to regulate designated sources of storm-water discharges); § 1345(d)(2)(A)(i) (deadline for EPA identification of toxic pollutants, whose presence in sewage sludge may adversely affect public health or the environment).
27. 7 U.S.C.A. § 136a-1 (setting deadlines for reregistration of pesticides' active ingredients).
28. Lujan v. Defenders of Wildlife, 504 U.S. 555 (1992).
29. Lujan v. Nat'l Wildlife Fed'n, 497 U.S. 871 (1990).
30. Lujan v. Defenders of Wildlife, 504 U.S. 555 (1992).
31. *Id.* at 566, 567.
32. *Id.* at 564.
33. Massachusetts v. EPA, 549 U.S. 497 (2007).
34. *Id.* at 541.
35. *Id.* at 516.
36. *Id.* (quoting Lujan v. Defenders of Wildlife, 504 U.S. at 580 (Kennedy, J., concurring in part and concurring in judgment).
37. *Id.*
38. *See* Bradford C. Mank, *Standing and Future Generations: Does Massachusetts v. EPA Open Standing for Generations to Come*, 34 Colum. J. Envtl. L. 1 (2009) (the Court's approach "would arguably allow Congress to confer standing on certain persons to act as representatives of future generations as long as the statute was clear and the plaintiffs had suffered injuries that also threatened future generations.").
39. Massachusetts v. EPA, 549 U.S. at 1453.
40. *Id.* at 1454–55 (reasoning that "[h]aving surrendered some of their sovereign abilities to protect their environments when they entered the union – for example, the ability to negotiate for greenhouse gas reductions with foreign powers – states were now reliant on Congress to help protect their 'quasi-sovereign interests.'").
41. *See* Cass Sunstein, *What's Standing after* Lujan*? Of Citizen Suits, "Injuries" and Article III*, 91 Mich. L. Rev. 163, 232–33 (1992); *see also* Myriam E. Gilles, *Reinventing Structural Reform Litigation: Deputizing Private Citizens in Enforcement of Civil Rights*, 100 Colum. L. Rev. 1384 (2000) (proposing same concept in the civil rights field).
42. 31 U.S.C. §§ 3729–31.
43. Vt. Agency of Natural Res. v. United States *ex rel.* Stevens, 529 U.S. 765 (2000).
44. *Id.* at 772.
45. *Id.*; *see also* Sprint Commc'ns Co. v. APCC Servs., Inc., 128 S.Ct. 2531 (2008) (holding that a company that is assigned a debt has standing to sue on behalf of the debtor, although the plaintiff will not recover any money as the result of a lawsuit).

46. Harold Feld, *Saving the Citizen Suit: The Effect of* Lujan v. Defenders of Wildlife *and the Role of Citizen Suits in Environmental Enforcement*, 19 COLUM. J. ENVTL. L. 141, 149 (1994).
47. *Stevens*, 529 U.S. at 772.
48. *Id.* at 773.

3 Shifting Baselines and Backsliding Benchmarks

The Need for the National Environmental Legacy Act to Address the Ecologies of Restoration, Resilience, and Reconciliation

Thomas T. Ankersen[1] and Kevin E. Regan[2]

A BASELINE IS A POINT OF DEPARTURE; A BENCHMARK, A POINT along a continuum that marks progress toward a goal. Central to the design of the National Environmental Legacy Act (Legacy Act or Act) is the collection of baseline data about natural resources, the definition of the nature and quality of the environmental legacy to be preserved, and ongoing monitoring to ensure preservation of the desired legacy. If the goal of the Legacy Act is to ensure the intergenerational transfer of a legacy, then a baseline must be set, benchmarks for success established, and current ecological realities confronted. None of these are simple tasks, and they have been made infinitely more complicated by the accelerated pace of anthropogenic change, which has resulted in both shifting and already-shifted baselines. We use the term "shifting baselines" to describe the phenomenon of how humans' perception of normal or baseline environmental conditions can shift dramatically over time, particularly between generations. The pitfall of shifting baselines is that environmental degradation often goes unrecognized by successive generations, which may not appreciate the degraded state of what they perceive as a pristine and functional ecosystem.

The shifting-baselines phenomenon poses several major challenges to designing an effective Legacy Act. First, if we fail to consider historical environmental conditions and set conservation goals based on already-shifted baselines, we may constrain – even doom – the resource legacy we seek to transfer. Second, resource management decisions that fail to address gradual environmental change may subtly shift baselines, diminish the environmental legacy, and eventually push ecosystems to the brink. Finally, accelerated anthropogenic change ensures that some baselines will shift irreversibly, and in some cases, new, transformed ecosystems will emerge and stabilize within foreseeable generations. Failure to understand and address these challenges in ways more sophisticated than the traditional metrics of species counts, acreage amounts, and pollutant levels could doom Legacy Act. To preserve the options available to future generations, an effective legacy act must consider the legacy of ecological functions, processes, services, and their interactions, as well as the ability of ecosystems to absorb and adapt to change. An effective legacy act must embrace what Hollings and others have described as the "governance of resilience."[3]

In this chapter, we first review natural science literature that discusses the shifting-baselines phenomenon, which has also been described by environmental psychologists as "environmental generational amnesia"[4] or the "extinction of experience."[5] We next examine some of the problems posed by the oversimplification of baselines through a discussion of ecological restoration and socio-ecological resilience theory. We suggest that an informed and effective legacy act must consider three facets of contemporary ecology: (1) restoration, (2) resilience, and (3) reconciliation – the reconciliation of humans and the ecosystems they inhabit (or used to inhabit). For many ecosystems, baselines have shifted to the extent that simply transferring our present environmental legacy cannot ensure that the legacy will be received by future generations fully endowed. In many instances, the restoration of degraded ecosystems will be required to ensure the preservation of a legacy of ecosystem functionality. Setting benchmarks that also envision the goal of restoring degraded ecosystems – in other words, the baseline of a previous generation – will help ensure the resilience of these ecosystems, which must be robust to adapt to an uncertain future. To ensure ecosystem resilience, many scientists and policy analysts believe that the best approach is to maximize biodiversity and its components. Thus, we believe that biodiversity represents the most appropriate overarching metric to gauge progress toward the goal of endowing a legacy that does not diminish over time. To further this goal, the precautionary principle should guide resource management decision making.

Finally, we suggest that features of the National Environmental Legacy Act that can help confront the challenges that the shifting-baselines phenomenon poses for complex ecological and social systems. Conserving, even restoring, ecosystems on public lands will not be sufficient. It will be necessary to reconcile human and nonhuman species' needs by establishing a mosaic of interconnected habitats and by reintegrating ecological processes and functions into human-dominated landscapes.

We conclude that the shifting-baselines phenomenon also requires resetting the baseline of traditional ecological knowledge; and this remains a necessary antecedent to the intergenerational transfer of an environmental legacy. It is necessary to increase levels of ecological understanding, including awareness of historical ecological conditions and how those conditions have been shaped by anthropogenic change, to foster the sociopolitical support for a fully endowed national environmental legacy.[6]

1. The Shifting-Baselines Phenomenon

The term "shifting baselines" has been attributed to Dr. Daniel Pauly, who first described the "shifting baseline syndrome" in the context of fisheries science.[7] He explained:

> Essentially this syndrome has arisen because each generation of fisheries scientists accepts as a baseline the stock size and species composition that occurred at the beginning of their careers, and uses this to evaluate changes. When the next generation starts its career, the stocks have further declined, but it is the stocks at that time that serve as a new baseline. The result obviously is a gradual shift of the baseline, a gradual accommodation of the creeping disappearance of resource species, and inappropriate reference points for evaluating economic losses resulting from overfishing, or for identifying targets for rehabilitation measures.[8]

As an example of this phenomenon, Pauly described a study based on historical accounts that estimated that the biomass of fish and other organisms used by humans along the North Atlantic coast of Canada is less than 10 percent of what it was two centuries ago. Pauly explained that some modern fisheries managers may have difficulty accepting that early fishing methods could have had such dramatic impacts, but he posited that large animals of low fecundity at the top of earlier food webs "must have been less resilient to fishing than the survivors that are exploited today."[9] Pauly recommended that fisheries science develop frameworks for incorporating historical knowledge, including information that some scientists might dismiss as anecdotal. He explained that the incorporation of historical accounts can provide a more complete context for the modern field of fisheries science, and thereby "help us to understand and to overcome – in part at least – the shifting baselines syndrome, and hence to evaluate the true social and ecological cost of fisheries."[10]

Since Pauly's initial description, numerous authors have elaborated on the concept of shifting baselines in a variety of disciplines. In particular, the phenomenon has been frequently discussed in marine science literature, which may reflect the fact that marine ecosystems and organisms have been among the most rapidly degraded and imperiled.[11] Indeed, the decline of coral reefs may best exemplify the shifting-baselines phenomenon.[12] Individuals who encounter coral reefs on a regular basis have been able to observe the decline of coral-reef ecosystems during their lifetimes, and in some instances, in a matter of decades. The shifting-baselines phenomenon is also particularly relevant to the field of marine science as a result of the long-held perception of the oceans as vast and of limitless resources.[13] It is only relatively recently that humans have begun to fully appreciate the fragility of marine environments and the decline of marine species.

Shortly after Pauly's initial publication, Charles Sheppard cautioned that, "[i]f baselines are shifting, then many ecological models which relate in some way to 'natural' conditions will have been programmed with erroneous starting points."[14] Sheppard posed the question: "What apparently clean coastal areas have not been affected by even modest inputs of sewage and increased land run-off, year by year, over the past 200 years?"[15] Another author, James Carlton, also discussing shifting baselines in the marine science context, explained, "Without a framework of study and a deeper appreciation for marine environmental history, our sense of history often defaults to viewing the step on which we are standing as the second step of the staircase, no matter how far down the staircase we have gone."[16] He cautioned: "Lest we think that fish, for example, were at peak abundances 'back when' – say[,] 100 years ago – we need to think again. One hundred years ago both fishermen and scientists remembered 'back when' as well."[17]

Subsequent studies have provided empirical support for the shifting-baselines phenomenon. One study examined the perceptions of fishermen in an area of Mexico's Gulf of California where fish populations had declined steeply over a sixty-year period.[18] Three generations of fishermen from Baja California were interviewed about the species and places that they considered to have been depleted by over-fishing. In addition, they were asked about their experiences with Gulf grouper (*Mycteroperca jordani*), including questions about the best catch they remembered landing, the largest animal they ever caught, and the year in which these catches were made. The results of this study indicated a rapid intergenerational shift in perception of the local marine environment. The study found that despite plentiful times of large fish still being within living memory, few young fishermen appreciated that large species had ever been common or that nearshore sites had ever been so productive.[19]

The study also observed that, although older fishermen recalled greater species abundance than the young fishermen did, the baselines observed by the older individuals had shifted from those that early European visitors observed. For example, few older fishermen commented on the once-valuable fishery for pearl oysters (*Pinctada mazatlanica*), which were abundant from the beginning of the seventeenth century until the fishery collapsed in 1940.[20] Similarly, few individuals were aware of the largest predators, such as the white shark (*Carcharodon carcharias*) or the goliath grouper (*Epinephelus itajara*) that once furnished meals for many of the seventeenth- and eighteenth-century buccaneers in the area.[21]

Another marine study examined the precipitous decline of pelagic (open ocean) sharks in the Gulf of Mexico by comparing catch rates between the 1950s and the late 1990s.[22] That study concluded that the oceanic whitetip shark (*Carcharhinus longimanus*) and silky shark (*C. falciformis*), which were formerly the most commonly caught shark species in the area, have declined by more than 99 percent and 90 percent, respectively.[23] That study noted that "recent papers on pelagic sharks have either not mentioned the oceanic whitetip or have dismissed it as a rare species, with no recognition of its former prevalence in the ecosystem."[24] It also emphasized the importance of "retrospective analyses" to understand the full magnitude and nature of human impacts. The study explained: "The perception of what was natural in the open ocean has clearly changed over a very short period (less than half a century), and our results suggest that it may be particularly easy for baselines of incidentally harvested species to shift because they are usually poorly monitored."[25] Thus, the study reflects the importance of systematic monitoring to develop a baseline and detect changes in that baseline.

Several other studies have attempted to estimate historical environmental conditions to more fully illustrate the dramatic decline of certain species. One study analyzed maps of historic sea turtle nesting areas to measure changes in distribution and abundance.[26] The results of that study indicated that 20 percent of the historical nesting sites had been lost entirely and 50 percent of the remaining nesting sites had been reduced to dangerously low populations. The study also concluded that, while recent conservation efforts have resulted in large population increases at several nesting sites, the loss of widespread nesting areas throughout the Caribbean and the reduction of the Caribbean-wide population since human hunting began indicate that Caribbean turtles are far from recovery.[27] That study observed that the reduction in sea turtle numbers affected marine ecosystems because of the important functional role of sea turtles, which modify the environment through their grazing activity and by metabolizing cellulose, which increases available nutrients in the food chain.[28]

Furthermore, two studies examined the impact of human activity on coral-reef ecosystems over various historical time frames. One study described the significant impact that human activities have had on coral reefs since the arrival of Spanish explorers, both in terms of the species composition and in terms of the function of the coral-reef ecosystem.[29] Another study observed that, "[c]ontrary to romantic notions of the oceans as the 'last frontier' and of the supposedly superior ecological wisdom of non-Western and precolonial societies, our analysis demonstrates that overfishing fundamentally altered coastal marine ecosystems during each of the cultural periods we examined."[30] That study also noted that "[c]hanges in ecosystem structure and function occurred as early as the later aboriginal and early colonial-stages, although these pale in comparison with subsequent events."[31] Thus, it appears that shifting baselines is not an exclusively modern phenomenon, and that anthropogenic impacts have shaped ecosystems over the course of centuries and even millennia.

In contrast to the research described previously, which examined the shifting-baselines phenomenon at the species level, one study analyzed general shifts in baselines of toxicity from an ecosystem level.[32] That study compared the toxicity of a highly publicized oil spill that occurred in 2002 off the coast of Spain with background toxicity levels in the area.[33] The analysis concluded that, compared to the daily increase in levels of overall toxicity over a period of decades, the oil spill was statistically "insignificant – a mere drop in a glass of water."[34] The author of that study, J. M. Ruiz, observed: "[T]his precisely is the tragedy, that such a serious accident is actually irrelevant from an eco-toxicological point of view."[35] Ruiz also discussed some of the sociopolitical implications of the shifting baselines phenomenon: "If society is led to believe that the marine environment is without problems – as long as there is no oil spill, then governments will be reinforced in the Pavlovian behaviour: the apparent risk merits major attention, the unnoticed one deserves little or none."[36] He argues that ecologists have a responsibility to communicate the insidious threat of gradual and systematic environmental degradation, including climate change, overfishing, eutrophication, biological invasion, and pollution.[37]

The concept of shifting baselines has also been explored in interdisciplinary contexts, including ecopsychology. James Miller, while discussing potential strategies for developing broad-based public support for biodiversity conservation, explores the intersection of the research of ecologists and ecological psychologists.[38] Miller discusses the relationships between the shifting-baselines phenomenon and what has alternatively been described by some authors as the "extinction of experience" or "environmental generational amnesia."[39] Miller emphasizes that "the environment encountered during childhood becomes the baseline against which environmental degradation is measured later in life," and he cautions that the result is a "continual ratcheting down of expectations regarding the quality and ecological function of natural areas closest to people's homes and workplaces."[40] Miller cites the work of the noted entomologist Robert Pyle, who used the term "extinction of experience" to describe human disconnection from the natural world and a gradual state of apathy toward natural systems.[41] Subsequently, this apathy results in more environmental degradation and thereby perpetuates a vicious cycle of separation and further isolation from nature.[42]

Empirical support for the shifting-baselines phenomenon in the context of human psychology can be found in the work of the environmental psychologist Peter Kahn. In one influential study, Kahn and Friedman interviewed inner-city African American children in a polluted area of Houston, Texas, about their environmental views and values.[43] The study found that approximately two-thirds of the children understood the concepts of air and water pollution, but only one-third of the children believed that those environmental issues affected them directly.[44] The authors of the study questioned how children who knew about pollution in general and lived in a polluted urban environment could be unaware of the pollution around them. Kahn and Friedman suggested one possible explanation: "if one's only experience is with a certain amount of pollution, then that amount becomes not pollution, but the norm against which more polluted states are measured."[45] They also noted that, "with each generation[,] the amount of environmental degradation increases, but each generation takes that amount as the norm – as the nonpolluted condition."[46] Kahn and Friedman described this trend as "generational amnesia."[47]

Kahn later elaborated on the concept of environmental generational amnesia and described the pitfalls of shifting baselines:

> The upside is that each generation starts afresh, unencumbered mentally by the environmental mistakes and misdeeds of previous generations. But the downside is enormous in that each of us can have difficulty understanding in a direct, experiential way that the nature as experienced in our childhood was already environmentally degraded. Thus, we're constructing our environmental ethic, and structuring our relationship with nature, based on incomplete and partly inaccurate perceptions and understandings.[48]

Thus, Kahn provides a description of the shifting-baselines phenomenon on the individual level. His work provides insight into the mechanism by which baseline perceptions can collectively shift over time. It also illustrates that it is ecological knowledge – especially personal ecological experience – that determines one's perception and expectations of the natural world.

2. Baseline Considerations: Restoration, Resilience, and Reconciliation

The shifting-baselines phenomenon suggests a number of considerations that are useful for defining the contours of the National Environmental

Legacy Act. These insights should affect how baselines are determined under the Act, how benchmarks are set, and how we define and measure the legacy that we wish to pass to future generations. First, rather than focusing solely on present conditions, it is necessary to consider historical environmental conditions when defining an environmental legacy. Professor Flournoy advocates public deliberation to determine the appropriate legacy. Such a public dialogue should be informed by a full understanding of ecological conditions, including historical context. Here, the field of restoration ecology offers guidance, as restoration ecologists continue to wrestle with the question of how to define success in ecosystem management in the context of shifted and sometimes unknowable baselines. Defining the legacy for the purposes of the Act must be informed both by historical environmental knowledge and by recognition that future anthropogenic change is inexorable.

Second, benchmarks – which can play an important role in shaping future baselines –must be built into the Legacy Act's architecture. Planning horizons must be stretched, and the possibility of ecological change already set in motion must be considered. As Professor Robert Deyle has stated when describing current shortsighted decisions concerning urban infrastructure and sea-level rise: "Development decisions being made today are committing public and private capital to land use patterns and associated infrastructure and facilities with design lives that reach well into the period of time when the impacts of sea level rise will be felt."[49] The same can be said for legacy decisions in the context of anthropogenic change. Land and resource use decisions are being made today that commit or constrain those resources to a present baseline condition that extends well into the period of time when the impacts of climate change will begin to take their toll on ecosystems. In some instances, it may be necessary to set benchmarks that attempt to restore ecosystems to previous historical conditions to ensure that healthy, or simply functional, ecosystems can actually be passed to the next generation. As is discussed further herein, restoration may be necessary to achieve ecosystem resilience to prevent the collapse of ecosystems that are already in jeopardy of imperilment and to allow ecosystems to adapt to future conditions.

However, anthropogenic change that has already taken place may impose limits on our ability to restore ecosystems to previous conditions. Climate change remains the proverbial elephant in the room for the Legacy

Act and all other forward-looking environmental thought. Irreversible processes have been set into motion that ensure that the resource legacy will not be the same from one generation to the next, and that the resource legacy certainly will not be in the same place. What good does it do to restore the freshwater Everglades when some models suggest that much of that region will become tidal within the century? There may be a good answer for this,[50] but if we do not consider it when defining the baseline legacy for this ecosystem and assigning benchmarks for the achievement of the intergenerational transfer of that legacy, we will we be doomed to disappointment.

Accordingly, it is necessary to take steps beyond maintaining the ecological status quo. Take, for example, our legacy of biodiversity: if we wish for this legacy to keep pace with climate change, it may not be enough to maintain our present spatial baseline of protected areas that serve as reservoirs of biodiversity; many will need to be expanded and linked to allow for species migration in response to temperature, rising seas, and other climate shifts.[51] In some cases, this will require protection of not only public resources, as the Legacy Act currently considers, but privately held resources as well. As is discussed herein, voluntary conservation and restoration efforts will be a necessary complement to the Act, and new approaches to, and scales of, land use regulation may be required.

Third, once environmental memory is lost, it becomes more difficult to pass an awareness of past conditions to the next generation, ultimately resulting in what has been called "environmental generational amnesia."[52] Thus, to secure an environmental legacy, it is necessary to foster environmental knowledge of the past, instill an awareness of how the environment is currently changing as a result of human impacts, and ignite the hope of how things could be. Contact and familiarity with the natural world allows individuals to perceive if the environment around them is changing. Humans' experiences with the natural world can shape their attitude toward the environment around them.[53] Thus, environmental education, and perhaps more important, ensuring human interaction with nature, is essential to avoid baselines shifting from the pitfalls of environmental generational amnesia. This has been described by some authors as an important tenet of "reconciliation ecology."[54] Stewardship agencies' Legacy Act mandate should include a duty to communicate information about the resources they manage with the public. In addition, maintaining or creating opportunities for the public to interact with natural systems should be one of the values considered and required to be preserved when making decisions that may affect the environmental legacy.

In this section of the chapter, we discuss considerations that should inform the process of defining an environmental legacy and setting a substantive standard under Legacy Act. These considerations include determining an appropriate measurement of the current environmental baseline and deciding what future benchmarks we wish to achieve. These benchmarks can, in turn, serve as baselines for the environmental legacies of future generations.

First, we discuss the concepts of restoration ecology and suggest that setting a current baseline and future benchmarks that reflect restoration goals is necessary to ensure a meaningful environmental legacy that includes functional and sustainable ecological systems. Second, we discuss the concept of ecological resilience, which describes the ability of an ecosystem to absorb and adapt to disturbance, and we suggest that baselines and benchmarks should be set at a level that provides for ecosystem resilience. Third, we discuss the difficulties of determining baselines and benchmarks in light of great uncertainty about how natural systems function, and we suggest that biodiversity may be the most useful metric of ecosystem health and function. Fourth, we suggest that concepts of reconciliation ecology, a field that strives to reconcile human dominated and natural systems, should inform Legacy Act to help ensure that the baseline of human ecological knowledge does not further diminish and that the appreciation of natural systems necessary to preserve and enhance an environmental legacy is not lost. These concepts also suggest the importance of creating mosaics of interconnected anthropogenic and natural landscapes, and of increasing biodiversity levels within human-dominated systems. Legacy Act must embrace these broader goals.

A. Lessons from Restoration Ecology

Restoration is defined as the "process of assisting the recovery of an ecosystem that has been degraded, damaged, or destroyed."[55] Restoration incorporates the idea of a historical baseline or a set of historical conditions, determined by a combination of knowledge of the system's preexisting structure, function, and services; comparisons with similar ecosystems that have suffered less severe damage; and information on regional environmental conditions.[56] Restoration is considered complete when the historical baseline or condition is achieved, and the restored ecosystem no longer requires human intervention to function as a healthy, vigorous, and sustainable system.[57]

The difficulties in establishing a historical baseline are common to restoration goal setting and to identifying shifting baselines.[58] Quite often ecologists and other decision makers lack adequate data – both quantitative and qualitative – to establish an appropriate or accurate baseline. When decision makers consider individual and collective memory (traditional ecological knowledge), they must determine what information is adequate and reliable. Restorationists must also choose a realistic and feasible restored condition.

James A. Harris argues that historical references and baselines may be less useful in creating direct objectives and more useful in giving context to restoration goals and developing models for future ecosystem formations.[59] For example, because of the scale and pace of anthropogenic change, restoration to a historic baseline may be impractical at best and impossible at worst. Even so, Harris recognizes that historical references are crucial to understanding the range of ecosystems that a location will support.[60]

Harris proposes focusing on restoration of the natural capital and ecosystem goods and services rather than solely focusing on "those metrics based on the numbers and arrangement of the biota."[61] He advances three concepts of ecological fidelity that should be considered when setting restoration goals: (1) structural or compositional replication, (2) functional success, and (3) durability.[62] Structural or compositional replication refers to the need for the restored ecosystem to resemble, as closely as possible, an appropriate reference ecosystem. Functional success means that the restored ecosystem should "align ecologically with the system it is designed to reproduce."[63] Finally, durability refers to the ability of the ecosystem to be sustained for a relatively significant period of time.[64]

The concepts of ecological fidelity that can guide restoration efforts reflect human perceptions of nature, wilderness, and the environment. William Throop argues that historical fidelity is a value-laden concept that extends beyond pure ecological terms.[65] Historical fidelity, he asserts, depends on human knowledge and understanding of the welfare of the ecosystem components, the duty of current generations to future generations, and human conceptions of the virtues expressed through restoration. Although historical fidelity, that is, the attempt to achieve historical conditions, is a compelling force in restoration, critical reexamination of its role may ultimately clarify goals for restorationists.[66] From a legacy perspective, restoration turns back the generational clock and sets the baseline to a prior generation's legacy – which has since shifted. If the goal of Legacy Act is simply to turn over to the next generation a resource stock that is not significantly diminished, then restoration may be viewed as a luxury, a gift to (or burden on) the future. However, it may be the case that restoring ecosystems to a prior generation's baseline is necessary to avoid significant diminution to the future estate, by restoring processes functions and services to avoid some future tipping point. Many ecosystems and species are currently pushed to the brink. Concepts of ecological resilience suggest that certain levels of biodiversity and ecosystem function are necessary to preserve the flexibility of ecosystems to adapt to future change. This flexibility is necessary to ensure the long-term survival of ecological diversity and ensure the transfer of an environmental legacy that ensures intergenerational equity.

B. Considerations of Ecological Resilience

Considerable attention has been devoted to seeking to understand the behavior of entire ecosystems in the face of change agents. One important conceptual framework for this discourse is known as resilience theory. Resilience is defined as "the capacity of a system to absorb disturbance and reorganize while undergoing change so as to retain essentially the same function, structure, identity and feedbacks."[67] Ecosystems that can no longer absorb disturbance are prone to what Folke and colleagues refer to as "regime shifts," the transformation of one ecosystem to another, less desirable one.[68] Often, these less desirable states include diminished capacity to deliver ecosystem goods and services, a key anthropocentric rationale for transferring the resource legacy to a succeeding generation. In some cases, the regime shift may be irreversible. Folke and colleagues argue that, in such cases, managing ecosystem transformation rather than restoration is appropriate; humans should concentrate on facilitating and then maintaining "a new stability landscape."[69] Where regime shifts are avoidable, resilience theory argues for a framework for avoidance that essentially keeps a system away from thresholds or tipping points for regime shift. Complete ecological restoration may not be necessary to accomplish this. But understanding when and where those tipping points may occur can form the basis for establishing restoration baselines. In the absence of knowledge of the tipping points, a precautionary approach is required. The Legacy Act incorporates these insights by making resilience an overriding goal and mandating that stewardship agencies monitor and manage ecosystems to maintain ecosystem resilience.

According to Folke and colleagues, resilience is lost and regime shifting results where there has been a removal of functional groups of species (e.g., whole trophic levels), through the emission of waste and pollutants (including those that contribute to global warming), and through alteration of the magnitude frequency and duration of disturbance regimes (e.g., fire, hurricanes), or a combination of these. To accomplish the Legacy Act's goals, the Act must consider, and where appropriate, regulate, each of the different types of effects to prevent tipping points and regime shifts that may limit the environmental legacy that is passed to future generations.

C. The Potential for Biodiversity to Serve as a Baseline and Benchmark Despite Uncertainty

In the face of socio-ecological complexity, there is some scholarly consensus on the role of biodiversity and the provision of environmental services both as a metric for the resilience of ecosystems and as the basis for a new environmental law regime. Biodiversity, which is defined as the "variation of life at all levels of biological organization,"[70] may serve as the most effective guiding star for setting legacy baselines and benchmarks. As Folke and colleagues note: "The diversity of functional groups in a dynamic ecosystem undergoing change, the diversity within species and populations, and the diversity of species in functional groups appear to be critical for resilience and the generation of ecosystem services."[71]

Environmental law scholars have also argued for the use of biodiversity as the basis for a new approach to addressing complexity in ecological systems. Professor J. B. Ruhl, who characterizes the contemporary environmental law framework as "reductionist" and linear, has called for a "revolutionized environmental law" based on considerations of policy, process, and performance.[72] Ruhl begins by offering the principle of "sustainable development," "or sustainability," as the central organizing principle for environmental law.[73] He then embraces the policy process of adaptive management as the means to achieve sustainability. Adaptive management is also a tool for implementing resilience theory.[74] Finally, Ruhl suggests that biodiversity, and the provision of ecosystem services, should serve as the performance standard for measuring sustainability.[75] Hence, there appears to be some convergence among resilience theorists and environmental law scholars that biodiversity offers the most appropriate measure of ecosystem sustainability.[76] We agree that biodiversity may also provide the best metric for establishing both baselines and benchmarks for the intergenerational transfer of the natural resource legacy.

One additional concept merits discussion when considering the role of baselines in formulating a law designed to ensure the intergenerational transfer of ecological wealth. The old adage "the more we know; the less we know" seems particularly apt for the ecological sciences. The venerable "balance of nature" has been supplanted by terms like "discordant harmonies,"[77] "biocomplexity,"[78] and "dynamic equilibria."[79] Underlying each of these postmodern bases of contemporary ecology is the acknowledgment that policy development must account for the scientific uncertainty embraced by these new visions of the natural world. A former U.S. secretary of defense famously articulated this premise in a rather different context when he said: "There are known knowns; there are things we know we know. We also know there are known unknowns; that is to say we know there are some things we do not know. But there are also unknown unknowns – the ones we don't know we don't know."[80] In the policy parlance of the environment, Donald Rumsfeld's "known unknowns" are known as "tipping points," "trophic cascades," and "synergistic effects."[81] These necessitate reliance on another underlying principle of the new environmental law, which, like the principle of sustainable development, originated outside the borders of the United States. The precautionary principle, or precautionary approach, as it is also called, provides a policy response to scientific uncertainty.[82] This principle, articulated in a number of international agreements, provides that, "[w]here there are threats of serious or irreversible damage, lack of full scientific certainty shall not be used as a reason for postponing cost-effective measures to prevent environmental degradation."[83] The complexity of ecological systems, particularly given the current levels of anthropogenic interference, virtually ensures that scientific uncertainty will beleaguer any effort to establish meaningful baselines and benchmarks. The precautionary principle suggests that natural resource legacy benchmarks – biodiversity benchmarks – should exceed, perhaps greatly exceed, what is expected to be required to accomplish an intergenerational transfer of the existing legacy.

D. Reconciliation Ecology and the Potential to Restore Ecological Knowledge

In addition to restoration ecology and resilience theory, another means to confront shifting baselines is reconciliation ecology.[84] We use the term "reconciliation ecology" to describe a form of conservation science that strives to modify the places dedicated to human activities to provide for greater biodiversity and ecosystem functionality in those places.[85] Reconciliation ecology can be especially useful for areas that have already been modified by human development and for which it is impractical to engage in full-scale restoration efforts. One 2004 empirical study illustrates the relationship between the shifting-baselines phenomenon and efforts to increase biodiversity in human-dominated landscapes.[86] Turner and colleagues compared biodiversity in five cities spanning three continents and calculated biodiversity levels within neighborhoods.[87] The study concluded that most residents in the study were concentrated in areas of impoverished biodiversity, and that this impoverishment was more pronounced in terms of native species.[88] The authors of that study noted the potential consequences of their findings in terms of the shifting-baselines phenomenon: "[i]f the baselines by which humans assess ecological health diminish as new generations are exposed to poor ecological conditions, the fact that the greatest numbers of people live below [mean neighborhood diversity] virtually guarantees the future decline of these baselines."[89] The authors noted: "Logically, there are two options for reducing the displacement of humans from biodiversity: Either move humans to nature, or bring nature to humans."[90] It is the latter option that is the focus of reconciliation ecology.[91] Michael Rosenzweig offers several species-specific examples of reconciliation ecology, such as constructed ponds that provide habitat for the natterjack toad (*Bufo calamita*) in the United Kingdom and the successful reproduction of

American crocodiles (*Crocodylus acutus*) in the cooling canals of a power plant in South Florida.[92]

One advantage of a reconciliation ecology approach is that it can engender consensus about conservation efforts and allow humans to interact with nature.[93] James Miller observes that reconciliation ecology offers a framework for developing a wider focus in conservation, that it can engage a larger group of stakeholders, and that it "moves beyond the dichotomy of pristine habitats and spoiled habitats."[94] For example, Miller cites the importance of smaller nature preserves in Kane County, Illinois, which, though they may not be large enough to sustain populations of grassland-obligate bird species, allow other taxa such as prairie-obligate butterflies to survive.[95] Miller also provides the example of urban nature reserves in the Chicago area, including a green roof on Chicago's city hall, and a network of natural areas that includes nearly 360,000 acres of protected lands and waters.[96] He explains that the value of these smaller ecological reserves should not be evaluated solely in traditional ecological metrics but also "should also be measured at the scale of human experience."[97] He explains: "By sheer virtue of their accessibility, these areas may do much to enhance the value of a prairie in the minds of those who live nearby. This in turn may translate into a more widespread recognition of the necessity of conserving more extensive grasslands elsewhere."[98] By encouraging human interaction with nature and by increasing awareness of the existing environment, as well as awareness of how that environment has changed, it may be possible to facilitate restoration of ecological knowledge.[99] Such an increase in ecological knowledge may help to avoid the shifting-baselines phenomenon by allowing humans to appreciate that the environment around them is in a degraded state and, in turn, by motivating them to support efforts to improve that environment.

The concept of reconciliation ecology encourages new collaborations among social scientists, designers, and planners, as well as "a broader appreciation for the connections between biodiversity conservation and quality of life in the places where most of us live."[100] For example, encouraging the establishment of native landscapes in suburban neighborhoods, with native hedges serving as microcorridors that lead to buffered urban streams, could help create coupled human and natural habitats that lead to larger, more intact ecosystems. Homeowners' associations in Florida that manage common areas for key species,[101] such as tortoises, provide another example of the application of reconciliation ecology.[102]

The concept of establishing a mosaic of interconnected anthropogenic and natural habitats is also consistent with the biosphere reserve concept promoted by the UN Educational Scientific and Cultural Organization's (UNESCO) Man and Biosphere Reserve Programme.[103] This model envisions the trilogy of core protected area, buffer zones, and transition zones.[104] In this three-tiered land use strategy, the inner preservation core is protected by separate buffer and transition zones that serve as sites for resource conservation and experiments in sustainable resource use and development.[105] The buffer areas permit limited levels of use without imposing the strict limitations of the protected area of the more human-dominated land uses of the general region.[106] However, reconciliation ecology suggests that even these human dominated systems must attempt to create opportunities for native species and humans to coexist.[107]

3. A Modest Proposal for the Legacy Act

A review of the literature of shifting baselines, restoration ecology, resilience theory, and reconciliation ecology suggests that the Legacy Act must confront significant ecological and social challenges, and embrace restoration, resilience, and reconciliation. Otherwise, a substantive standard that maintains the status quo by prohibiting only actions that "substantially diminish the quantity or quality of public natural resources" could allow baselines to continue to shift, ecosystems to flip, and legacies to be lost. Accordingly, we recommend that the Act confront the ecological and social dimensions of shifting baselines.

First, given the currently degraded state of many ecosystems, section 1 of the Act, which establishes the Act's goals and policies, must explicitly address the long-term goal of restoration by stating that the purpose of the Act is not only to maintain the present resource legacy for future generations but also to enhance that legacy through restoration. At a minimum, the purpose of the Act must contain a broadly worded purpose similar to that of the Clean Water Act, which states that its purpose is "to maintain and *restore* the physical, chemical and biological integrity of the nation's waters."[108] A findings provision of the statute should recognize that ecosystem change, including global climate change, has occurred at accelerated rates from anthropogenic impacts and that a comprehensive statutory framework is necessary to ensure the continuation of ecosystem processes and functions. This provision should emphasize that the most reliable metric for ensuring these processes and functions is biodiversity.

Second, the determination of the Act's substantive standard and establishing a process for revisiting that standard is critical. The phenomenon of shifting baselines illustrates the difficulty of setting a substantive standard. Scientific literature demonstrates that humans have had significant impacts on the natural world for centuries and that even preindustrial societies shaped ecosystem processes and functions.[109] In light of this natural and anthropogenic ecosystem evolution, it is not clear which long-term benchmarks should be set. However, the concept of ecosystem resilience, measured by the capacity of an ecosystem to endure exogenous shocks, may provide some forward momentum if it can be operationalized statutorily.

Section 8, which addresses the problem of impacts to resources that are either not the result of human conduct or result from preenactment human conduct, is a critical component of the Act. In her chapter, Professor Flournoy suggests that stewardship agencies must conduct resilience assessments to determine whether ecosystems can persist in the face of potential ecosystem changes. She notes that, "[i]f the assessment suggests that the ecosystem cannot persist – that it has flipped or inevitably will flip into a new behavior regime without intervention – the statute does not assume that restoration should always be undertaken." Arguably, the precautionary principle suggests that doubts should be resolved in favor of some type of restoration or an attempt at restoration. However, as she notes, in some instances, restoration might be both costly and fruitless, while in others, an ecosystem might be able to be restored and its values and services might warrant the cost of doing so. We agree with her suggestion, that, at a minimum, the decision of whether to seek to restore an ecosystem should be the subject of at least a notice and comment proceeding to permit public input and debate of the value choices as well as the technical questions. However, principles of adaptive learning also suggest that stewardship agencies must periodically reconsider decisions about whether to engage in restoration efforts to account for new information, changed circumstances, or new technology.

Third, the Act must also address the social dimensions of shifting baselines. Environmental ethics and environmental education play an important role in shaping the goals that society chooses to set. The benchmarks that are achieved or not achieved in one generation dictate the baseline of the next generation. The literature on the psychological dimension of shifting baselines demonstrates that humans' experiences, particularly in childhood, determine how they perceive the natural world and the environmental ethic that they internalize. Only by increasing ecological knowledge and awareness of ecosystem change can the social and political support necessary to avoid shifting baselines and backsliding benchmarks be achieved. When agencies consider the values and services provided by legacy resources, the value provided by resources in increasing ecological knowledge should be explicitly recognized.

Monitoring is essential to detecting baselines that are shifting. Thus, section 5 of the Act is critical. However, it is vital that these data be used to inform decisions and facilitate adaptive learning. Stewardship agencies must not only collect these data but also communicate the significance of the data to the public. Only by informing the public about changes that are occurring in ecosystems, some of which may not be readily observable, will the social and political consensus to change existing resource management decisions be possible.

Finally, the Act must consider the impact that activities, including land use or land cover change, on private property has on the public resources that are the subject of the Act. Professor Flournoy's concept proposal should consider a provision that recognizes this central point and directs greater federal participation in local land use decisions that may affect legacy transfers. The federal government has generally shied away from land use matters. However, if we accept that the public lands may not be large enough to ensure ecosystem resilience in many cases, then some framework for consideration of the federal interest in land use should be developed. One approach might be a regulatory or participatory overlay based on the concept of buffer zones and transitions zones used in planning protected areas.

4. Conclusion

The shifting-baselines phenomenon poses an insidious threat to the Legacy Act's goal to ensure that we can provide future generations with an environmental legacy that is equal to or greater to that which we received.

Maintaining and transferring the status quo on public lands may not be sufficient to ensure intergenerational natural resource equity. For the Legacy Act to be effective, its substantive standard must explicitly incorporate restoration of already-degraded resources, a legacy the generation has received. The Legacy Act must set the baseline for the current generation's resource transfer at a level sufficient to ensure ecosystem resiliency in the face of reasonably anticipated anthropogenic change. The precautionary approach must be the polestar used to provide for resilient ecosystems. The act should adopt the metrics of biodiversity as most appropriate baseline and benchmark setting and monitoring metrics. We also contend that the Legacy Act will not fully succeed unless it recognizes the role that private property must play in ensuring the integrity of ecological landscapes. This may require a more explicit federal participation in local land use decision making. Nor will the Act succeed unless it comes to terms with the intergenerational ecological amnesia that is responsible for shifted and shifting baselines by creating a robust program to restore traditional ecological knowledge among ordinary Americans.

NOTES

1. Legal Skills Professor & Director, Conservation Clinic, University of Florida Levin College of Law.
2. Associate Attorney, Earthjustice, a nonprofit public interest law firm. The views expressed in this Chapter that Mr. Regan coauthored are his own and no endorsement of Earthjustice or its clients is intended.
3. Carl Folke, Steve Carpenter, Brian Walker, Marten Scheffer, Thomas Elmqvist, Lance Gunderson & C.S. Holling, *Regime Shifts, Resilience, and Biodiversity in Ecosystem Management*, 35 Ann. Rev. Ecology, Evolution & Systematics 557–81 (2004).
4. James R. Miller, *Restoration, Reconciliation, and Reconnecting with Nature Nearby*, 127 Biological Conservation 356, 357 (2006).
5. James R. Miller, 20 *Biodiversity Conservation and the Extinction of Experience* Trends Ecology & Evolution 430, 431 (Aug. 2005).
6. *See generally* Richard Louv, Last Child in the Woods: Saving Our Children from Nature-Deficit Disorder (2006).
7. Daniel Pauly, *Anecdotes and the Shifting Baseline Syndrome of Fisheries*, 10 Trends Ecology & Evolution 430 (1995).
8. *Id.*
9. *Id.*
10. *Id.*

11. *See* Nancy Knowlton & Jeremy B. C. Jackson, *Shifting Baselines, Local Impacts, and Global Change on Coral Reefs*, 6(2) PLOS Biology (2008) (noting that most of the world's tropical oceans are heavily degraded and that most modern (post-SCUBA technology) ecological studies of coral-reef ecosystems have focused on ecosystems that are moderately to severely degraded).
12. *See* Mary Gray Davidson, *Protecting Coral Reefs: The Principal National and International Legal Instruments*, 26 Harv. Envtl. L. Rev. 499, 504 (2002) (describing the rapid decline of coral-reef ecosystems).
13. *See* James Carlton, 12(6) *Apostrophe to the Ocean*, Conservation Biology 1165, 1166 (Dec. 1998) (noting "[T]he oceans, unlike forests, still look like the oceans after we've removed their contents").
14. Charles Sheppard, *The Shifting Baseline Syndrome*, 30(12) Marine Pollution Bulletin, 766 (1995).
15. *Id.*
16. Carlton, *supra* note 13, at 1166.
17. *Id.*
18. *See generally* Andrea Sáenz-Arroyo et al., *Rapidly Shifting Environmental Baselines among Fishers of the Gulf of California*, 272 Proc. Royal Soc'y Bulletin 1957 (August 2005).
19. *Id.* at 1961. That study concluded that, compared to young fishermen, older fishermen named five times as many species and four times as many fishing sites that had once had abundant fish but had since been depleted. The older fishermen also caught up to twenty-five times as many Gulf grouper as young fishermen on their best fishing day ever. Although 96 percent of older fishermen and 90 percent of middle-aged fishermen had caught a Gulf grouper, only 45 percent of young fishermen had. *Id.* at 1960. Statistical analysis demonstrated that the biggest Gulf grouper were disappearing and that the results were not an artifact of older fishermen being more skilled or having fished for longer. *Id.* at 1960. The average size of the largest fish ever caught by the older fishermen, estimated from a length-weight relationship, was 84 kilograms (184.8 pounds), that of the middle-aged fishermen was 72 kilograms (158.4 pounds), and that of the younger fishermen was 63 kilograms (138.6 pounds). *Id.* at 1960. Although many older fishermen remembered abundant medium-sized sharks, large groupers, large snappers, and sea turtles, the middle-aged fishermen were unaware of the past abundances of these animals. *Id.* at 1959. Indeed, few young fishermen were aware that such species had ever been common. *Id.*
20. *Id.* at 1959.
21. *Id.*
22. *See generally* Julia K. Baum and Ransom A. Myers, *Shifting Baselines and the Decline of Pelagic Sharks in the Gulf of Mexico*, 7 Ecology Letters 135 (2004).
23. *Id.* at 142.
24. *Id.*
25. *Id.* at 143.

26. Loren McClenachan et al., *Conservation Implications of Historic Sea Turtle Nesting Beach Loss*, 4(6) FRONTIERS ECOLOGY & ENVIRONMENT 290 (2006).
27. *Id.* at 293–94.
28. *Id.* at 292.
29. Jeremy B. C. Jackson, 16(Supp.) *Reefs since Columbus*, CORAL REEFS S23–S32 (Jan. 1997).
30. Jeremy B. C. Jackson et al., *Historical Overfishing and the Recent Collapse of Coastal Ecosystems*, 293 SCIENCE 629, 636 (July 2001).
31. *Id.* at 636.
32. *See generally* J.M. Ruiz, *Oil Spills versus Shifting Baselines*, 282 MARINE ECOLOGY PROGRESS SERIES 307 (Nov. 16, 2004).
33. *See generally id.*
34. *Id.* at 308.
35. *Id.*
36. *Id.*
37. *Id.*
38. *See generally* Miller, *supra* note 5.
39. Miller, *supra* note 4, at 357.
40. Miller, *supra* note 5, at 431.
41. *Id.* (citing R.M. Pyle, *The Extinction of Experience*, 56 HORTICULTURE 64, 64–65; ROBERT M. PYLE, THE THUNDER TREE: LESSONS FROM AN URBAN WILDLAND (Houghton Mifflin 1993)).
42. *Id.*
43. Peter H. Kahn Jr. & Batya Friedman, *Environmental Views and Values of Children in an Inner-City Black Community*, 66 CHILD DEV. 1403 (1995).
44. Peter H. Kahn Jr., *The Child's Environmental Amnesia – It's Ours*; 17(2) CHILDREN, YOUTH & ENVT'S 199, 203 (2007) (summarizing the results of Kahn and Friedman's 1995 study).
45. Kahn & Friedman, *supra* note 43, at 1414.
46. *Id.*
47. *Id.*
48. Kahn, *supra* note 44, at 204.
49. Robert E. Deyle, Katherine C. Bailey & Anthony Matheny, ADAPTIVE RESPONSE PLANNING TO SEA LEVEL RISE IN FLORIDA AND IMPLICATION FOR COMPREHENSIVE AND PUBLIC FACILITIES PLANNING, FLORIDA PLANNING AND DEVELOPMENT LAB, Department of Urban and Regional Planning, unpublished report submitted to Florida Department of Community Affairs (2007).
50. *See* Cornelia Dean, *The Preservation Predicament*, N.Y. TIMES, Jan. 29, 2008. Restoration advocates contend that the restoration goal of increasing freshwater flows south toward Everglades National Park will help to maintain a hydrostatic head that will hold back the rising sea and allow the accumulation of peat to serve as a barrier to further inward migration of the estuarine ecosystem. *See also Climate Change Testimony to Congress Subcomm. before the Subcomm. on*

Interior, Environment, and Related Agencies of the H. Appropriations Comm. Concerning Climate Change and Lands Administered by the Department of the Interior, Apr. 26, 2007 (statement of Dan Kimball, superintendent, Everglades National Park) (describing potential effects of climate changes and actions to mitigate those effects).

51. *See generally* Adrian D. Manning et al., *Stretch Goals and Backcasting: Approaches for Overcoming Barriers to Large-Scale Ecological Restoration*, 14(4) RESTORATION ECOLOGY 487 (Dec. 2006); Janne Bengstsson et al., *Reserves, Resilience and Dynamic Landscapes*, 32(6) AMBIO 389 (Sept. 2003).
52. Miller, *supra* note 4, at 357.
53. One author notes: [W]hat is the extinction of the condor to a child who has never seen a wren?" ROBERT M. PYLE, THE THUNDER TREE (Houghton Mifflin 1993).
54. Miller, *supra* note 4, at 359. Miller explains that the goal of "reconciliation ecology" is to reconcile human needs with those of native species by designing human habitats to meet the habitat requirements of native species. *Id.*
55. James A. Harris et al., *Ecological Restoration and Global Climate Change*, 14 RESTORATION ECOLOGY 170, 172 (2006). Restoration differs from, and often precedes, management; although they may overlap with respect to the required human interactions, management is intended to guarantee the health of the system after it has been restored.
56. Society for Ecological Restoration International Science & Policy Working Group, *The SER International Primer on Ecological Restoration* (Oct. 2004), *available at* http://www.ser.org/content/ecological_restoration_primer.asp.
57. *Id.*
58. One author notes: "Restoration projects that seek to restore a bay, an estuary, or a marsh to the way it looked when the proponents were young – as if this temporal target was by default the aboriginal world – seek a world that was already a barely detectable shadow of its former self, and would be better served by a more rigorous temporal target." Carlton, *supra* note 13, at 1167.
59. Harris, *supra* note 55, at 173.
60. *Id.*
61. *Id.*
62. Eric S. Higgs, *What Is Good Ecological Restoration?* 11 CONSERVATION BIOLOGY 338 (1997).
63. Eric S. Higgs, NATURE BY DESIGN: PEOPLE, NATURAL PROCESSES, ECOLOGICAL RESTORATION 128 (Mass. Inst. Tech. Press 2003).
64. *Id.*
65. *Available at* http://eco.confex.com/eco/2007/techprogram/P3338.HTM.
66. HIGGS, *supra* note 63, at 143–48.
67. Brian S. Walker, C.S. Holling, S.R. Carpenter & A. Kinzig, *Resilience, Adaptability and Transformability in Social-Ecological Systems*, 9(2) ECOLOGY & SOC'Y 5 (2004), *available at* http://www.ecologyandsociety.org/vol9/iss2/art5/ (last accessed June 23, 2009).

68. Carl Folke et al., *supra* note 3.
69. *Id.*
70. Kevin J. Gaston & John I. Spicer, Biodiversity: An Introduction (Blackwell, 2d. ed. 2004).
71. Folke et al., *supra* note 1 (citing Chapin et al. 1997 & Luck et al. 2003). "Functional groups of species in a system refers to groups of organisms that pollinate, predate, fix nitrogen, spread seeds, decompose, generate soils, modify water flows, open up patches for reorganization, and contribute to the colonization of such patches. The persistence of functional groups contributes to the performance of ecosystems and the services that they generate." *Id.* These authors further break down the notion of functional groups into "biological legacy groups" and "mobile link species" and their supporting landscapes and seascapes. Legacy groups support reordering of the ecosystem after disturbance, whereas link species connect habitat over time and through space.
72. J.B. Ruhl, *Thinking of Environmental Law as a Complex Adaptive System: How to Clean Up the Environment by Making a Mess of Environmental Law*, 34 Houston L. Rev. 933 (1997).
73. *Id.* at 996–1000. Many people argue that sustainable development has emerged to become a principle of customary international law, modifying the unbridled right to development enshrined in the UN Charter. *See* Gabcikovo-Nagymaros Dam (Hung. v. Slovk.), 1997 I.C.J., 36 I.L.M. 700 (Weermantry, J., concurring).
74. *See id.* at 996–97.
75. Ruhl, *supra* note 72, at 997–99.
76. This convergence is even more noteworthy given that Ruhl, writing in 1997, did not draw from ecological resilience theory, then in its infancy, as the underlying science to support his environmental law revolution.
77. Daniel Botkin, Discordant Harmonies: A New Ecology for the 21st Century (Oxford University Press 1997).
78. Madhur Anand & Brian Tucker, *Defining Biocomplexity: An Ecological Perspective*, 8(4–5) Comments on Theoretical Biology 497–510 (2003); William K. Michener, Thomas J. Baerwald, Penelope Firth, Margaret A. Palmer, James L. Rosenberger, Elizabeth A. Sandlin & Herman Zimmerman, *Defining and Unraveling Biocomplexity*, 51(12) BioSci. 1018–23 (Dec. 2001).
79. See Richard O. Brooks, Ross Jones & Ross A. Virginia, Law and Ecology: The Rise of the Ecosystem Regime (Ashgate 2002).
80. *See* http://politicalhumor.about.com/cs/quotethis/a/rumsfeldquotes.htm (Donald Rumsfeld, Defense Department Briefing, February 12, 2002). Despite the derision with which it was greeted, Rumsfeld's terminology was not original and had already been in use in discussions of uncertainty and game theory, as well as risk management and decision theory. See http://en.wikipedia.org/wiki/Known_unknown#cite_note-2.
81. Ecological "unknown unknowns" are even more problematic. *See* Michener et al., *supra* note 78, at 1019–20. Michener and colleagues describe the characteristics of

biocomplexity, which refers to "properties emerging from the interplay of behavioral, biological, chemical, physical, and social interactions that affect, sustain, or are modified by living organisms, including humans." *Id.* at 1018. They point out that biocomplexity may be reflected in nonlinear, chaotic, or even unpredictable behaviors, and that "it may be typified by interactions that are likely to span multiple hierarchical levels as well as several spatial and temporal scales." *Id.* at 1019–20.

82. Timothy O'Riordon & James Cameron, INTERPRETING THE PRECAUTIONARY PRINCIPLE (Earthscan 1994).
83. Conference on Env't & Dev., *Rio Declaration on Environment and Development*, principle 15, UN. Doc. A/CONF.151/26 (Vol. I)(August 12, 1992).
84. *See* Miller, *supra* note 4, at 357.
85. Miller, *supra* note 5, at 433 (citing M.L. Rosenzweig, *Loss of Speciation Rate Will Impoverish Future Diversity*, 98 PROC. NAT'L ACAD. SCI. USA 5403–10 (2001); M.L. ROSENZWEIG, WIN-WIN ECOLOGY: HOW THE EARTH'S SPECIES CAN SURVIVE IN THE MIDST OF HUMAN ENTERPRISE (Oxford University Press 2003).
86. Will R. Turner, Toshihiko Nakamura & Marco Dinetti, *Global Urbanization and the Separation of Humans from Nature*, 54(6) BIOSCI. 585–87 (June 2004). Turner et al. provide empirical data that shows that, in cities worldwide, most residents are concentrated in neighborhoods of impoverished diversity, notwithstanding the fact that substantial biodiversity in the cities overall. *See generally id.*
87. *Id.* at 585–87.
88. *Id.* at 587.
89. *Id.* at 588 (internal citations omitted).
90. *Id.*
91. *See generally* Miller, *supra* note 5; Robert R. Dunn et al., *The Pigeon Paradox: Dependence of Global Conservation on Urban Nature*, 20(6) CONSERVATION BIOLOGY 1814–16 (2006).
92. Miller, *supra* note 5, at 433.
93. Miller, *supra* note 5, at 432, notes that "conservation strategies that emphasize quality-of-life enhancement might be more effective than appeals from environmentalists for altruism or self-sacrifice."
94. *Id.* at 433.
95. Miller, *supra* note 4, at 360.
96. Miller, *supra* note 5, at 432–33; *see also* http://www.chicagowilderness.org (last accessed June 23, 2009).
97. Miller, *supra* note 4, at 360.
98. *Id.*
99. Indeed, it is the interactions between humans and nature that can allow humans to forge a connection with, and come to appreciate, the natural world. Research indicates that children who play in wild environments show a greater affinity and appreciation for such places later in life. Miller, *supra* note 5, at 431.
100. *Id.*

101. "Key" species are species that are "important to ecosystem structure and function in whatever form (e.g., biomass, abundance, productivity, or functional role), driving ecosystem process or energy flows. Karen A. Bjorndal & Alan B. Bolten, *From Ghosts to Key Species: Restoring Sea Turtle Populations to Fulfill their Ecological Roles*, Marine Turtle Newsletter No. 100 (2003) at page 19 (citing S. Piraino, S. Fanelli, & F. Boero, *Variability of Species' Roles in Marine Communities: Change of Paradigms for Conservation Priorities*. Marine Biology 140: 1067–74 (2002). Bjorndal and Bolten note that sea turtles were "once key species in marine ecosystems," and that "the decline of sea turtles and other macroinvertebrates initiated the collapse of marine ecosystems in which they live." *Id.* at 19.
102. See http://www.ashtonbiodiversity.org/pdf/fs_homeowners_assoc.pdf and http://www.ashtonbiodiversity.org/tortoisereserve.php (last accessed June 23, 2009).
103. Thomas T. Ankersen, Kevin E. Regan & Steven A. Mack, *Towards a Bioregional Approach to Tropical Forest Conservation: Costa Rica's Greater Osa Bioregion*, 38 FUTURES 406–31, 410 (2006) (citing H. Hinote, *Framework for Integrated Ecosystem Management: The Southern Appalachian Man and Biosphere Cooperative*, in ECOSYSTEM MANAGEMENT FOR SUSTAINABILITY: PRINCIPLES AND PRACTICES ILLUSTRATED BY A REGIONAL BIOSPHERE RESERVE COOPERATIVE at 88 (J.D Pein ed., Lewis 1999). Ankersen et al. examine the concept of bioregionalism and the biosphere reserve model in the context of tropical forest conservation. However, these notions may also have applicability in different bioregions in the United States.
104. *Id.* (citing M. Batisse, *Developing and Focusing the Biosphere Reserve Concept*, 22(3) NATURE & RESOURCES 2–11 (1986)).
105. *Id.* (citing W.D. Solecki, *Putting the Biosphere Reserve Concept into Practice: Some Evidence of Impacts in Rural Communities in the United States*, 21(2) ENVTL. CONSERVATION 242 (1994)).
106. *Id.* (citing K. Garratt, *The Relationship between Adjacent Lands and Protected Areas: Issues of Concern for the Protected Area Manager*, in NATIONAL PARKS, CONSERVATION AND DEVELOPMENT: THE ROLE OF PROTECTED AREAS IN SUSTAINING SOCIETY (J. McNeely & K. Miller eds., Smithsonian Inst. Press 1984)).
107. *See generally* Miller, *supra* note 4, at 357.
108. 33 U.S.C. § 1251(a) (emphasis added).
109. Jackson et al., *supra* note 30, at 636.

4 Valuing Nature

The Challenge of the National Environmental Legacy Act

Mark T. Brown and Mary Jane Angelo

THE PROPOSED NATIONAL ENVIRONMENTAL LEGACY ACT (LEGACY Act or Act) provides an opportunity to achieve sustainable natural resource management. Changing our goals and aspirations for the future to reflect a more sustainable relationship with the environment requires that we look beyond the small scale, self-interest and the short term to the larger scale, the greater good, and the longer term. Important to implementation of the proposed statute is a mechanism to systematically evaluate whether actions affecting public resources will impermissibly degrade or deplete resources over the legacy period. The statute requires the identification of metrics to assess the impacts of proposed actions on a wide array of resources. This chapter proposes embodied-energy (emergy) synthesis as a tool that may provide a useful measure of values embodied in resources that may help to provide a readily quantified metric for use in setting resource baselines and assessing whether impacts will impermissibly degrade or deplete resources.

Emergy accounting is a donor system of value (measuring intrinsic value) based on solar energy required to produce things. This chapter briefly introduces the concepts of emergy synthesis and emergy accounting and describes how these could be employed to perform many important responsibilities under the Legacy Act. It then describes the challenges presented by the Legacy Act related to assessing the values and services provided by public natural resources. The chapter concludes with a more detailed explanation of the underlying methodology and theory of emergy accounting and the advantages of emergy accounting over available alternatives.

Emergy Synthesis: An Overview

In this chapter, we propose a conceptual framework, emergy synthesis,[1] and a procedure for environmental accounting consistent with it that stems from the intrinsic value of natural resources and is quantified as the solar energy required to make them. Termed "emergy synthesis" instead of emergy analysis, the conceptual framework is synthetic (i.e., puts together holistically) rather than analytic (i.e., breaks apart). Because emergy synthesis is predicated on a holistic view of economic goods and services, natural capital, and environmental services in the same framework, it is possible to apply an accounting approach that includes both the values of environment and the values associated with the economy. Termed "emergy accounting," the procedure relies on evaluating natural capital, environmental services, and economic goods and services without resorting to valuing techniques that ask for people's preferences. The measure of value called "emergy" and its monetary derivative "emdollars" is suggested as a central analytic tool that will facilitate implementation of the Legacy Act goals.

The origins of the emergy concept[2] date back to the 1950s, when Dr. Howard T. Odum, along with his brother Eugene, defined the crucial role of energetics in ecology.[3] Although he recognized the importance of energy in the early 1950s, it was in his book *Environment Power and Society* (1971) that Odum first formally touched on the concept of energy quality when he stated that it takes billions of acres of sunshine over millions of years to produce the concentrated power contained in fossil fuels.[4]

The concept of energy and resource quality evolved from the early 1970s, when it was primarily a qualitative description of different forms of energy, to a quantitative method for expressing concentration, flexibility, transportability, and convertibility of different forms of energy and resources relative to a common basis, the solar energy required to make them.[5] Identifying a need for a common denominator that could express different forms of energy and resources in the same framework, Odum developed the concept of emergy to describe the amount of one kind of energy required to make another form of energy or resource. At first the concept was called embodied energy, but because there were several definitions of embodied energy, Odum adopted the term “emergy” to set his concept apart and to capture the idea of energy memory,[6] or the quantification of the memory of energy of one form that is used up to make an energy or resource of another form.

From the 1970s through the present, the theory and concepts of emergy synthesis and the emergy accounting methodology have been further developed and refined by Odum, his students, and numerous other researchers throughout the world.[7] Emergy accounting methodology values resources on the basis of their intrinsic value rather than their monetary value (defined by neoclassical economics). Unlike neoclassical economic approaches that assign a monetary value to a resource on the basis of the price consumers are willing to pay to receive the value, emergy accounting is a donor value system. Emergy is based on the principle that the energy that goes into making a resource determines its value.[8]

Emergy Accounting

A formal definition of emergy is as follows: emergy is the availability of energy of one kind that is used up in transformations directly and indirectly to make a product or service. The unit of emergy is the “emjoule,” a unit referring to the available energy of one kind consumed in transformations. For example, sunlight, fuel, electricity, and human service can be put on a common basis by expressing them all in the emjoules of solar energy required to produce each. In this case, the value is a unit of solar emergy expressed in solar emjoules (abbreviated seJ).

Emergy accounting is the measurement and provision of information about environmental and human economic systems that can be used by decision makers to make resource allocation decisions and reasoned choices among alternative courses of action. Where financial accounting is the systematic measurement and reporting of financial information about a financial institution into standard forms for decision making,[9] emergy accounting is the measurement and provision of information about the interface of environmental and human economic systems useful to managers, regulators, and other stakeholders to make resource allocation decisions.

Emergy accounting evaluates all types of resources in the same units of solar emjoules, including natural resources (sometimes called natural capital, e.g., forest biomass, fish, soils, wildlife), renewable energy flows (e.g., sunlight, wind, rain), nonrenewable energy flows (e.g., fossil fuels), economic goods (e.g., finished products), labor, and information. In emergy accounting, both the monetary and the energy flows of the activity must be combined in the balance sheet to allow a balance for assets and liabilities to be determined. For ease of understanding by policy makers, managers, and other stakeholders, emergy units can be converted to monetary units. Using a combined emergy-monetary unit, called the emdollar, emergy accounting can be used to produce a single income statement and balance sheet giving comprehensive accounts for the economy, society, and the environment simultaneously.[10] The emdollar (^{em}\$) provides the means to relate emergy to money.[11]

Emergy accounting can be thought of as double-entry bookkeeping, in which the assets and liabilities of a particular situation (e.g., a given area of land, economic endeavor, or restoration project) are recorded in common units of emergy. Economic flows are converted to emergy units using a ratio of emergy to money. In traditional accounting procedures, the financial position or solvency of an organization, measured in monetary terms, is recorded using the fundamental equation of accounting:

$$\text{Assets} = \text{liabilities} + \text{equity}, \tag{1}$$

where assets are defined as economic resources, liabilities are economic obligations or debts, and equity is the remaining wealth after debts have been paid. The emergy balance sheet, for evaluating natural resource use, is a balance of assets and liabilities (or use). The equation is as follows:

$$\text{Current equity} = \text{current assets} - \text{liabilities}. \tag{2}$$

Current assets are divided into two categories: income from renewable environmental driving energies (e.g., sunlight, wind, rain, tides) and long-term environmental storages (resources) (e.g., forest biomass, wildlife species, soils, water). Liabilities are also listed under two categories: use of renewable environmental driving energies (e.g., use of rain for crop irrigation, use of wind for generation of electricity) and use of environmental storages (e.g., extraction of resources such as trees, fish, or other wildlife, or erosion of soil). Current equity is the difference between total assets and total liabilities and can be compared from year to year to determine net changes. The annual net change, if negative, can be considered an environmental deficit.

General Methodology for Emergy Accounting

The general methodology for emergy accounting is a top-down systems approach.[12] The first step is to construct systems diagrams that are a means of organizing thinking about relationships between components and pathways of exchange and resource flow. The second step is to construct emergy accounting tables directly from the diagrams. The final step involves calculating emergy indices that summarize and relate emergy flows of the economy with those of the environment. A further step-by-step elaboration of the methods used in emergy accounting follows.

Step 1: Overview System Diagrams

A system diagram in overview is drawn first to put the system of interest in perspective, to combine information from various sources about the system, and to organize data-gathering efforts. The process of diagramming the system of interest in overview ensures that all driving energies and interactions are included. Because the diagram includes both the economy and the environment of the system, it is like an impact diagram that shows all relevant interactions. The system diagram is used as a guide to construct a table of data requirements for the emergy accounting. Each pathway that crosses the system boundary is evaluated, as is each component in the systems.

Step 2: Emergy Synthesis Tables of Flows

Emergy evaluations of systems are carried out by accounting for the flows of materials, energy, and information that support the system. An emergy accounting table with the main headings and organization shown in Table 4.1 is used to organize data and maintain consistency with other evaluations. The evaluation of the system of interest must account for all the flows of material, energy, information, and money that support the system on an annual basis.

Step 3: Emergy Synthesis Tables of Storages

The storages of natural and economic capital are evaluated for the system of interest. The emergy synthesis table (Table 4.1) also contains line items for

Table 4.1. *Example of an emergy evaluation table: Emergy evaluation of one hectare of forested wetland in Florida*

Item	Data	Units	Emergy/ Unit (seJ/unit)	Solar Emergy[a] (E+15 SeJ)	em$ Value[b] (2005 US$)
Energy sources					
Sun	4.19E+13	J/ha/yr	1	0.04	22
Wind	3.15E+09	J/ha/yr	1,496	0.00	3
Rain, chemical potential	6.42E+10	J/ha/yr	18,199	1.17	622
Run-in, chemical potential	5.14E+10	J/ha/yr	90,995	4.67	2,487
Geologic input	5.50E+05	g/ha/yr	1.00E+09	0.55	293
Functions (Environmental services)					
Transpiration (water use)	2.57E+10	J/ha/yr	50,553	1.30	691
Gross Primary Production (GPP)	1.54E+12	J/ha/yr	2,377	3.67	1,953
Infiltration	3.61E+10	J/ha/yr	50,553	1.82	970
Structure (natural capital)					
Live biomass	2.79E+12	J/ha	65,914	183.58	97,650
Peat	3.78E+12	J/ha	160,474	605.82	322,244
Water	2.92E+10	J/ha	50,553	1.48	786
Basin structure	2.44E+07	J/ha	1.77E+11	4,308.66	2,291,840

[a] Solar emergy in this column is obtained by multiplying data in column 2 by the emergy/unit in column 4.

[b] Emdollar value in this column is obtained by dividing emergy in column 5 by 1.88 E12 seJ/$.

the main storages of the example system. The column headings in the storage table are the same as those in the flows table. To calculate the emergy of stored quantities (storages) of environmental resources – for instance, wood biomass or soils – it is necessary to sum the emergy of all of the inputs and then multiply by the time it takes to accumulate the storage. To calculate the emergy of economic storages – such as buildings and roads – all the inputs of energy, materials, and labor required to produce them are summed.

Step 4: Calculation of Emergy Metrics of Change

Metrics of change include the net change in assets, net change in annual productivity, net change in overall ecosystem health, and so on. The metrics are calculated from the flows of emergy-supporting processes and products and are used to provide perspective in comparisons with other processes and products, and pre- or post-alteration. Expected change over the legacy period can be evaluated. Emergy metrics can include ratios that relate economic emergy to environmental emergy, the intensity of emergy use, a ratio of benefit to cost, and a ratio that provides an index of environmental stress among others.

How Emergy Synthesis Supports Implementation of the Legacy Act

The National Environmental Legacy Act would require maintenance of a defined environmental legacy. Metrics[13] are needed both to define this legacy and to measure whether resources are being affected in such a way as to degrade or deplete this legacy. The Act contemplates systematic accounting of the impact of our decision making on the quality and quantity of our resources, now and in the future.

The Legacy Act demands that agencies (1) assess quantitatively and qualitatively the current state of public natural resources and their related services and values, and (2) ensure that decisions affecting those resources do not produce depletion or degradation that exceeds statutory limits. The goal of the statute is to assess the impact of decision making on the quality and quantity of our resources now and in the future. To permit such assessment and evaluation will require the development of metrics that capture or approximate relevant values and services. The challenge is how to express the value in a way that can be embraced by the political economy. Like the National Environmental Policy Act (NEPA), the Legacy Act requires the impacts of actions that affect the environment be taken into consideration before the actions actually begin, but unlike NEPA, the Legacy Act also requires agencies to evaluate in hindsight the losses resulting from interventions to adaptively learn from their actions. The very act of evaluating impacts demands an approach at the intersection of economy and environment. On the one hand, costs that can be measured in the economic realm have markets and subjective values that result in monetary prices. On the other hand, other costs affect values for which there are no markets. These values are intrinsic, result from the work of nature, and can result in common property goods and services with no price.

In keeping with Senator Gaylord Nelson's suggestion that "[t]he economy is a wholly-owned subsidiary of the environment,"[14] the proposed methodology assumes that all goods and services, whether from natural or technological systems, can be evaluated in the same overall accounting system that begins with the driving energies of the biosphere.

The conceptual framework of emergy synthesis can provide an accounting methodology that will inform decision making under the Legacy Act in three ways. First, emergy accounting can be used to provide quantitative baseline data against which future changes can be measured. Second, emergy accounting can be used to quantitatively evaluate the extent to which proposed action would result in depletion or degradation of the national environmental legacy. Finally, emergy accounting provides a tool for quantitatively evaluating data from monitoring of permitted actions.

Emergy accounting is unique among existing methodologies for valuing resources in that it inherently addresses the Legacy Act's two anchoring concepts, sustainability and concern for future generations. The objective of the Act is to define and protect a legacy of public natural resources for future generations. The Legacy Act's goal will be accomplished by setting forth a maximum level of degradation or depletion of resources that will be permitted over the course of the legacy period (i.e., a permissible level of degradation or depletion). Section 5 of the Act directs each stewardship agency to develop implementation regulations that designate appropriate metrics of quality and quantity for the resources for which they are stewards. Section 6 directs each stewardship agency to develop rules that set forth the quantity and quality of degradation or depletion of the relevant resource that constitutes impermissible degradation or depletion using the metrics developed under section 5. Section 6 of the Act limits the stewardship agencies' discretion to ensure that impermissible degradation or depletion does not occur. Emergy accounting could easily serve as the metric required to fulfill the requirements of these Legacy Act provisions.

Section 10 provides for exceptions to the prohibition on impermissible degradation or depletion of natural resources. An exception to the prohibition would apply where it can be shown by clear and convincing evidence that foreseeable technical advances or the availability of substitute resources obviate the need for or value of the resources in question. Another exception applies where impermissible degradation or depletion is clearly in the public interest, no acceptable alternative that will not cause impermissible degradation or depletion exists that will serve the public interest adequately, and the impacts to all services and values to be impaired can and will be mitigated. The conceptual framework of emergy synthesis could aid in making both of these determinations. First, emergy accounting could be employed to compare the quantity and quality of other substitute resources to determine whether such substitutes will, in fact, be adequate substitutes for the resource to be depleted. Second, emergy accounting can be used both to evaluate the impacts of alternatives and of proposed mitigation plans to determine whether they adequately offset both the quantity and quality of the degradation or depletion of natural resources.

Finally, section 8 of the Act authorizes citizen enforcement to seek penalties for any action that impermissibly degrades or depletes public natural resources. Emergy accounting could serve as a means for measuring the severity of degradation or depletion. Moreover, the extent to which degradation has occurred, as measured by means of emergy accounting, could provide useful information in determining an appropriate penalty or in fashioning other remedies in such enforcement actions. In addition, section 9 of the Act requires monitoring and adaptive learning. The emergy synthesis framework could be employed to evaluate information gathered through monitoring and to determine how to respond to such information in an adaptive manner.

We propose emergy synthesis as a value framework and emergy accounting as a bookkeeping methodology that could fulfill the purposes of the Legacy Act and be readily employed by all stewardship agencies to value the resources within their purview; to determine the extent of resource degradation and depletion; to evaluate information gleaned from long-term monitoring; and to determine whether alternative resources, including those proposed through mitigation, adequately offset the degradation or depletion proposed. To determine whether natural resource degradation is permissible under the Act, it is imperative to have a methodology that can assess the degree of degradation. To carry out the purpose of the Legacy Act, it will be necessary to employ a concise methodology to measure levels of natural resources degradation or depletion resulting from past actions or likely to result from future proposed actions. It must be able to capture the full range of natural resources value, including differences in the quality of resources and the replacement value if such resources are lost or degraded. Moreover, the methodology employed must be able to assess the quality and quantity of resources over a long period of time (the legacy period). The proposed Legacy Act's focus on sustainability and intergenerational equity is readily addressed in the framework of emergy synthesis, and the system of emergy accounting provides the tools for comprehensive intergenerational solvency (the ability to pay both economic and environmental debts).[15]

Emergy synthesis provides a number of significant advantages over other valuation methodologies, which directly address the concerns central to the Legacy Act. First, as described in greater detail herein, in contrast to the neoclassical economic analysis traditionally used in environmental decision making, emergy synthesis does not rely on the dollar value that humans assign to a particular resource but instead seeks to find the intrinsic value of the resources from an objective, scientific calculation of the amount of energy embodied in resources.

Second, unlike other methods of valuation, emergy synthesis explicitly addresses the difference in the quality of the energy embodied in particular resources.[16] "Quality," like "entropy" and "sustainability," is an ambiguous word, often translated into very elusive concepts. It may not have the same meaning to everybody, as the concept of quality is most frequently linked to human expectations. This might be called a receiver perception of quality. In the emergy synthesis framework, "quality" is defined as donor based rather than receiver based. By "donor-based quality," we mean that the donor instead of the receiver determines the value. This means that as the amount invested in something increases, its quality increases. From a donor basis, quality increases with increased inputs, such that the greater the expenditures of energy to make something, the better is its quality. In contrast, receiver quality is a subjective appreciation of the properties of an item. Although these subjective properties may reflect the needs of individual decisions and consumer behavior, subjective, qualitative, user-based values lack an objective basis from which public policy decisions can be made. In this larger context, a definition of quality is needed that is somewhat more than subjective. To achieve this, a change of scale is needed, shifting upward from the individual appreciation of a property to consider the role of the resource in the dynamics of the larger system in which it is embedded. By this shift, not only is the trap of a very subjective evaluation avoided, but another trap is escaped as well, that of market value. Something is valuable in the market if it is scarce compared to demand, no matter how much it contributes to the larger system in which it is traded. It is very likely that when a food is scarce, its market price (economic measure of quality) is high, in the very moment that its contribution to those who depend on it for life is lowest.

In contrast to emergy, economic valuation methodologies merely assign a dollar value to resources based on consumers' theoretical or actual willingness to pay, with little regard to the inherent quality of energy embodied in those resources. By accounting for energy quality as well as quantity, emergy accounting, in essence, provides a way to weight resources beyond looking at mere quantity. For instance, quality applies to animal and plant species just as it applies to different manufactured products, and the more energy that is invested in an animal (i.e., the higher it is on the food chain), the better is its quality. Thus, to capture ecosystem values in an emergy accounting framework based on species richness, the individual species that constitute the ecosystem are to be given weights based on their quality. Therefore, emergy synthesis provides a tool that can aid agencies that seek to assess the overall ecosystem impacts of decisions and actions.

Third, another benefit of using the emergy synthesis methodology is that it inherently recognizes the long-term value of resources. In contrast to neoclassical economic valuation methodologies that discount future values of resources in favor of their use today, emergy synthesis recognizes that not only do resources retain their value in the future but also their value can increase as they become scarcer. So-called discounting suggests that a resource user recognizes that monetary inflation will reduce the value of money received in the future and therefore wishes to receive funds now because future payments will be worth less. Emergy synthesis recognizes the same intrinsic value of resources now and in the future, as work potential does not depend on human preferences and is not affected by inflationary tendencies in the economy. In some instances (especially related to species), as a resource becomes scarcer, its emergy value increases. This is known as negative discounting: as species come close to extinction, their emergy values increase because of the significant energy requirements to re-evolve them from their nearest cousins.

Although the Legacy Act, as currently conceived, would leave it to each stewardship agency to develop its own methods of quantifying resources and metrics to evaluate change, we believe that it would be extremely beneficial if all stewardship agencies used a consistent framework for metric development that has the ability to evaluate all types of natural resources. Rather than each stewardship agency developing its own methods and metrics, potentially resulting in inconsistencies, confusion, and results that are not readily comparable, a common framework would ensure that each agency's data could be used by other agencies. In addition, a common framework would ensure that the public is better able to predict whether a particular proposed action is likely to result in impermissible degradation. Finally, a common framework will facilitate public oversight and enforcement roles with regard to stewardship agencies across the spectrum of federal agency responsibilities.

Emergy Synthesis: Energy Quality a Key Concept

In the foregoing, we provided a brief overview of emergy. Although it is beyond the scope of this chapter to entirely explain the thermodynamic basis and conceptual framework for emergy synthesis, in this section, we provide a brief explanation of one key principle, *energy quality*.

To explain the notion of energy quality, it is convenient to use the concept of an energy hierarchy (Figure 4.1) to visualize the work of nature and society that results in energy transformations. When viewed in totality, the systems of nature and society are interconnected in webs of energy flow, yet all energy transformations of the biosphere can be arranged in an ordered series to form an energy hierarchy with many joules of sunlight required to make a joule of organic matter, many joules of organic matter required to make a joule of fossil fuel, several joules of fuel required to make a joule of electric power, and so on.

Energy quality is calculated as the ratio of the emergy required to make something to its energy content. The term for this energy quality ratio is "transformity." A formal definition is as follows: Transformity is defined as the emergy input per unit of available energy output. For example, if

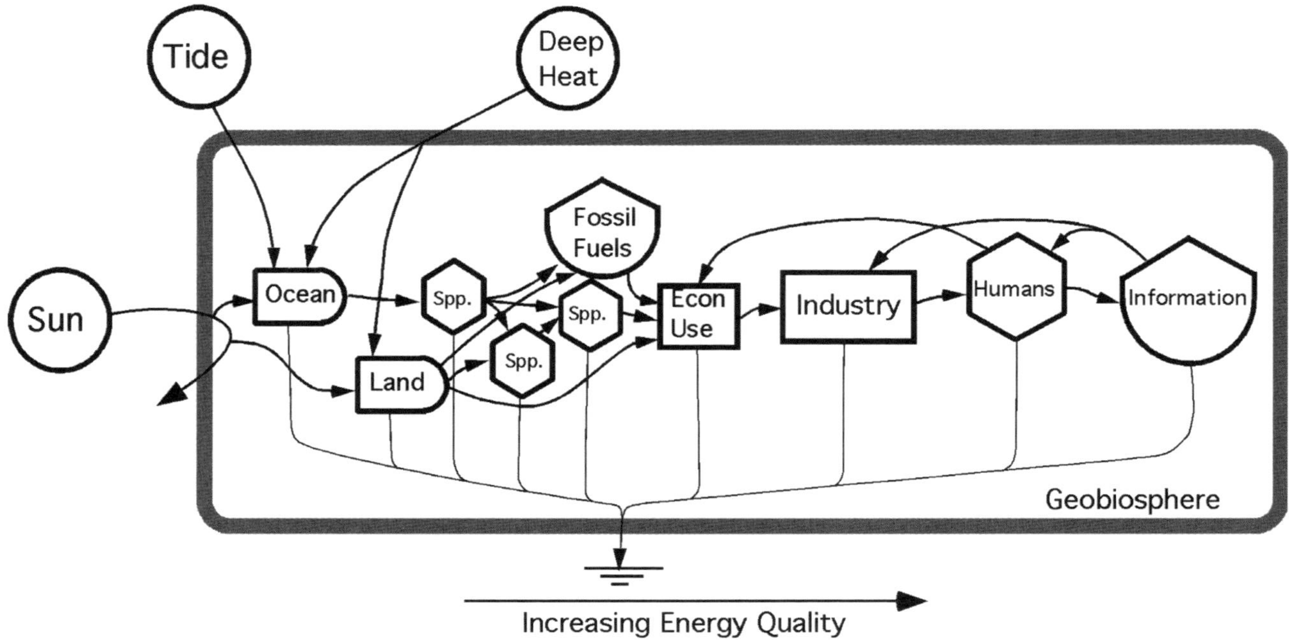

Figure 4.1. Aggregated systems diagram of the geobiosphere illustrating the concept of energy quality. At each energy transformation, from left to right, energies are upgraded to different forms that are more flexible, concentrated, and of higher quality.

four thousand solar emjoules are required to generate one joule of wood (i.e., the energy inputs from the sun needed to grow an amount of wood that when burned would release one joule of energy), then the solar transformity of that wood is four thousand solar emjoules per joule (abbreviated seJ/J). Solar energy is the largest but most dispersed energy input to Earth. The solar transformity of the sunlight absorbed by the Earth is 1.0, by definition.[17]

In Figure 4.1, the quality of energy increases from left to right as energies are transformed and upgraded (transformities increase). Thus, one joule of work by humans has better quality (higher transformity) than one joule of fossil fuels. Many studies have been completed over the years and transformities calculated for a variety of environmental services and components of natural capital.[18] Table 4.2 lists typical transformities for many environmental flows and assets of ecosystems. These flows and assets represent typical energy quality values for environmental services and natural capital of ecological systems. Transformities increase as the complexity of the systems and organisms increase. Table 4.3 lists typical transformities for organisms at various trophic levels in Prince William Sound, Alaska, and average transformities for trophic levels in Silver Springs, Florida, are given in Table 4.4 as examples of increasing quality of organisms in different ecological systems.

Use of Transformities in Emergy Accounting

Transformities are used in an emergy accounting procedure to convert quantitative evaluation of resource flows and storages to units of solar emergy for comparative purposes. The accounting procedure may be undertaken to evaluate some change in the environment; for this purpose, flows and storages are evaluated before and after the change. The changes (positive or negative) in flows and storages are quantified from observed data and then multiplied by their transformity to obtain the emergy, and total change (the algebraic sum of all changes) is computed. The emdollar (^{em}\$) value of flows and changes in storage reserves are computed from the emergy using a standard conversion factor. The emdollar values are more easily communicated to policy makers and the general public because they have units that are easily understood in most human value frameworks.

Table 4.2. *Summary of transformities in terrestrial ecosystems (after Brown and Ulgiati, 2004)*

Ecosystem	Transformity (seJ/J)
Gross primary production	
Subtropical mixed hardwood forest, Florida	1.03E+03
Subtropical forest, Florida	1.13E+03
Tropical dry savanna, Venezuela	3.15E+03
Salt marsh, Florida	3.56E+03
Subtropical depressional forested wetland, Florida	7.04E+03
Subtropical shrub-scrub wetland, Florida	7.14E+03
Subtropical herbaceous wetland, Florida	7.24E+03
Floodplain forest, Florida	9.16E+03
Net primary production	
Subtropical mixed hardwood forest, Florida	2.59E+03
Subtropical forest, Florida	2.84E+03
Temperate forest, North Carolina (*Quercus* spp.)	7.88E+03
Tropical dry savanna, Venezuela	1.67E+04
Subtropical shrub-scrub wetland, Florida	4.05E+04
Subtropical depressional forested wetland, Florida	5.29E+04
Subtropical herbaceous wetland, Florida	6.19E+04
Biomass	
Subtropical mixed hardwood forest, Florida	9.23E+03
Salt marsh, Florida	1.17E+04
Tropical dry savanna, Venezuela	1.77E+04
Subtropical forest, Florida	1.79E+04
Tropical mangrove, Ecuador	2.47E+04
Subtropical shrub-scrub wetland, Florida	6.91E+04
Subtropical depressional forested wetland, Florida	7.32E+04
Subtropical herbaceous wetland, Florida	7.34E+04
Wood	
Boreal silviculture, Sweden (*Picea abies*, *Pinus sylvestris*)	8.27E+03
Subtropical silviculture, Florida (*Pinus elliottii*)	9.78E+03
Subtropical plantation, Florida (*Eucalyptus* and *Melaleuca* spp.)	1.89E+04
Temperate forest, North Carolina (*Quercus* spp.)	2.68E+04
Peat	
Salt marsh, Florida	5.89E+03
Subtropical depressional forested wetland	2.52E+05
Subtropical shrub-scrub wetland	2.87E+05
Subtropical wetland	3.09E+05

Brown, M.T., and S. Ulgiati, 2004. "*Emergy, transformity and ecosystem health.*" In S.E. Jorgensen et.al. (eds.). Handbook of Ecological Indicators for Assessment of Ecosystem Health. Elsevier, New York.

Table 4.3. *Summary of transformities in a marine ecosystem: Prince William Sound, Alaska (after Brown et al., 1993)*

Item	Transformity (seJ/J)
Phytoplankton	1.84E+04
Zooplankton	1.68E+05
Small nekton (molluscans, arthropods, small fishes)	1.84E+06
Small nekton predators (fish)	1.63E+07
Mammals (seal, porpoise, beluga whale)	6.42E+07
Apex predators (orca)	2.85E+08

Brown. M.T., R.D. Woithe, C.L. Montague, H.T. Odum, and E.C. Odum. 1993. Emergy Analysis Perspectives of the Exxon Valdez Oil Spill in Prince William Sound, Alaska. Final Report to the Cousteau Society. Center for Wetlands, University of Florida, Gainesville, FL 114 pp.

In general, emergy evaluations of systems are conducted using average annual flows. For renewable inputs to the system of interest, such as solar energy, tidal energy, and rainfall, annual averages are used. For economic inputs, the most recent annual data available for the system of interest are used. Thus, although averaged flows are used for the renewable inputs, the data for economic flows are not averaged but consist of annual flows for a recent fiscal year. The result is an emergy evaluation of the system of interest, calculated as though it were in a steady state during the most recent year for which data are available.

Table 4.4. *Solar transformities of ecosystem components of the Silver Springs, Florida*

Item	Transformity (seJ/J)
Solar energy	1
Kinetic energy of spring flow	7,170
Gross plant production	1,620
Net plant production	4,660
Detritus	6,600
Herbivores	127,000
Carnivores	4,090,000
Top carnivores	40,600,000

Monetary Equivalents of Emergy

Because emergy is an unfamiliar concept and solar emjoules are more or less unknown in the policy arena, emergy values are often converted to monetary equivalents. Such conversions foster a better understanding of the relative values computed in emergy terms. The monetary units are equivalent to the currency of national economies in which the evaluations are conducted; however, to differentiate values derived from an emergy evaluation from those derived from an economic evaluation, the monetary unit is given the prefix "em." Thus, in the U.S. economy, computed emergy values are converted to emdollars. The emdollar (^{em}\$) value of resources, environmental service, or natural capital is calculated by first determining its emergy and then converting to emdollars using a standard conversion factor. The conversion factor is obtained by dividing the total emergy driving an economy by the economy's gross domestic product (GDP). Obviously, this is an average value for an entire economy and can be used only to provide perspective, suggesting that, on average, for every dollar circulating in the economy there is so much emergy driving it. Total emergy driving the economy in the United States (in 2005) was 1.89 E25 seJ and the GDP in that year was \$9.8 E12; thus, the ratio of emergy to dollars was as follows:

$$\text{seJ}/\$ = \frac{1.89\text{E}25\text{seJ}}{\$9.8\text{E}12} = 1.9\text{E}12\text{seJ}/\$. \quad (3)$$

Because the ratio of emergy to money is derived from the total emergy driving the circulation of money (GDP), it represents an index of monetary value of emergy. The ratio of emergy to money can be used to convert emergy values to a more familiar metric, dollars. Thus, an emergy quantity of, say, 3.8 E18 seJ is equivalent to ^{em}\$2 million (3.8 E18seJ/1.9 E12 seJ/\$). Table 4.5 lists emdollar values for examples of natural capital and environmental services obtained from a recent analysis of the U.S. Forest Service.[19]

Benefits of Emergy Synthesis

Numerous methodological approaches exist that can be used to evaluate natural resources under the Legacy Act. The predominant approaches to valuing resources, such as ecosystems services valuation, use neoclassical

Table 4.5. *Examples of the emdollar value of natural capital and environmental services*

Item	Units	(seJ/unit)	^{em}\$ (2005)
Natural capital			
Aquifer water	m^3	1.6E+12	0.85
Tree biomass	m^3	2.9E+14	155
Topsoil	m^3	3.1E+14	162
Herbivore	individual	4.6E+14	244
Carnivore	individual	4.0E+15	2,100
Top carnivore	individual	7.3E+16	38,200
Endangered species	individual species	4.0E+24	2.08E+12
Environmental services			
Net primary production	ha/yr	1.2E+13	6.5
Carbon sink	ha/yr	3.1E+13	16.2
Soil building	ha/yr	4.9E+13	25.9
Water supply	ha/yr	1.0E+15	546

economic approaches to assign a dollar value to natural resources. In addition, in recent years, a number of sustainability indices have been developed to measure environmental health and human welfare. However, these indices are of limited value, as they tend to view environmental health and human welfare only at a very broad (national) scale, which does not provide a means for evaluating the impact of changes made at a more localized level. In addition, a number of biodiversity indices exist. The limitation of these indices is that they typically value only species richness or species diversity, both of which are not necessarily good measures of ecosystem health or performance. Many approaches in the domain of ecological economics, though more comprehensive because they include factors not always included in more traditional economic approaches, are still grounded in neoclassical economics and therefore share many of the same shortcomings as other neoclassical economic approaches. Emergy accounting is a better Legacy Act metric because it avoids the problems of neoclassical economics and can be used to evaluate entire systems (including both natural and human components) at any scale using one common denominator.

Although neoclassical economic approaches remain the predominant approaches to valuing natural resources, such approaches have numerous shortcomings, which have been described at length in the scholarly literature.[20] One of the most significant shortcomings is the artificiality of placing a dollar value on natural resources using neoclassical economic methods. Although some natural resources, such as timber and minerals, have a market value, many natural resources are not bought and sold on the market and thus do not have a market value.[21]

Neoclassical economics addresses this problem by using contingent valuation to determine consumers' willingness to pay for a resource. The primary failing of contingent valuation is its underlying premise that natural resources have value only to the extent that consumers are willing to pay to preserve them.[22] This premise is inherently flawed because it assumes that consumers have perfect information and adequate technical fluency to be able to determine how much they would be willing to pay for a particular natural resource. The benefits of many natural resources are very complex and not readily apparent to the average layperson. For example, most lay consumers probably would not be willing to pay much to preserve a particular species of bacterium if they do not have complete information and technical understanding that the bacterium is critical to decomposition necessary to keep an ecosystem functioning. Moreover, studies demonstrate that willingness to pay, which is typically used in contingent valuation, is inherently skewed toward valuing the right to use resources rather than the right to preserve resources.[23] To protect a resource, consumers are generally willing to pay only about half the amount that the same consumers would be willing to accept to allow the resource to be exploited.[24]

At a more basic level, however, a fundamental problem with contingent valuation is that, from a scientific perspective, consumer preference has no relationship to the importance of the ecological resource or service for sustaining life on Earth.[25] Many ecological goods and services are not assigned any value by neoclassical economic analysis, and thus they are rarely included in any meaningful way in traditional economic valuation. Scientists have been working to develop alternative methods for assigning a value to ecological resources and services for many years.[26] Emergy synthesis is one such alternative valuation methodologies that provides a valuation methodology that relies on science rather than consumer preferences.

In addition to the fact that emergy accounting does not rely on contingent valuation, it has a number of other benefits over neoclassical economic systems because it is based on the principle that the energy embodied in a resource or service determines its value. In other words, emergy accounting seeks to discern the intrinsic value of the resource. Thus, emergy accounting is a donor system of valuation, as opposed to the receiver system of value on which neoclassical economics is based. This donor system of value based on solar energy required to produce resources rejects the underlying assumption of neoclassical economics valuation that value stems only from utilization by humans.[27] Emergy accounting has a number of benefits over other approaches to valuation: (1) it provides a bridge that connects economic and ecological systems, (2) it allows the quantification and comparison of the economic and ecological values of any system independent of monetary perception, (3) it overcomes the inability of money to value nonmarket resources, (4) it is scientifically sound and shares the rigor of thermodynamic methods, (5) it uses a common unit that allows all resources to be compared on a fair basis, (6) it recognizes the different qualities of energy, and (7) it provides a more holistic alternative to many existing methods for environmental decision making.[28]

A number of studies have demonstrated the value of emergy accounting as a method for evaluating environmental impacts and for comparing alternative development options to determine which option results in the least overall harm. For example, one study that evaluated three alternative sources of water supply for Windhoek, Namibia, demonstrated that the use of aquifer water was the preferable alternative, primarily because of the environmental and economic costs of desalination and the downstream environmental impacts to the Okavango Delta wetlands and wildlife should water from the Okavango River be diverted.[29] Another study compared impacts from tourism developments and selected the alternative with lowest impact by matching the intensity of economic development with the ability of the environment to absorb its environmental "load."[30] In yet another study, the environmental impacts of alternative hydroelectric dam proposals on the Mekong River were evaluated including the impacts to downstream ecosystems and fisheries. The net emergy of the dam configurations were calculated on the basis of the emergy costs versus the emergy yield.[31]

In addition to neoclassical techniques of analysis, many researchers have developed evaluation tools such as embodied energy analysis, exergy analysis, material flow accounting, life-cycle assessment, ecological footprint, and societal metabolism, among others. Each method is able to answer specific questions about a system's performance. Although some common features with emergy accounting can be found in some of these different methods, emergy accounting was intended to account for aspects that are usually not accounted for by other evaluation methods. Most nonemergy approaches evaluate only nonrenewable resources and do not account for the free services that a system receives from the environment, which are important requirements of productive process, often equally important as fossil fuels. In addition, many evaluation procedures do not have an accounting procedure for human labor, societal services, and information. Emergy accounting includes all of these and, in so doing, provides a powerful and comprehensive tool for understanding the dynamic interaction between human-dominated processes and resources and services provided for free by nature.

Although emergy accounting has many benefits over other methodologies of evaluating environmental costs and benefits, it should not be viewed as a panacea. The emergy value of a resource in itself does not necessarily provide absolute information about whether a resource is beneficial; it merely conveys that the resource possesses a certain level of embodied energy and therefore would require that level of emergy to replace. It also suggests that, if released in the environment, a resource with a high level of emergy will affect the environment in a way that is somewhat proportional to the level of emergy. Determination of whether the effect is positive or negative is a matter of public perception and becomes the domain of public policy debate. For example, DDT, a pesticide made from fossil fuels that requires a large amount of energy to make, has a relatively high emergy level. Because emergy-dense substances can have large environmental impacts, DDT has the potential to have significant impacts in ecosystems. It is now perfectly clear that it not only can control insect populations but also bioaccumulates in food chains, causing serious impacts on higher-order animals. Whether society is willing to accept the negative impacts when compared to the positive benefits is a matter of public policy.

One of the anchoring concepts of the Legacy Act is ensuring that our resources are preserved for future generations. To ensure such a legacy, it is necessary to recognize as a value today the value that a protected resource will have in the future. Neoclassical economic approaches to valuing resources do not recognize the future value of protecting resources. Instead, by applying a discount rate when valuing the future benefits that may result from protecting a resource, neoclassical approaches actually devalue the protection of resources for the future.

Although all neoclassical economic approaches use discounting, there is significant debate over the appropriate discount rates that should be used. The higher the discount rate that is used, the more likely it is that the analysis will conclude that the proposed resource degradation or depletion is insignificant. In contrast, a low discount rate assigns greater value to future benefits and, as a result, is more likely to support the conclusion that the benefits of resource protection are significant. The consequences of environmental decisions extend deep into the future. The further out in time we attempt to predict environmental consequences, the more the accuracy of such projections diminishes.[32] Using a low, or even no, discount rate is considered a way to "take the long view," because a low discount rate does not devalue long-term environmental benefits.[33] Discounting inevitably encourages a short-term view and hence biases public policy decision making. Moreover, our limited understanding of environmental issues provides further justification for avoiding the undervaluation of future benefits. Perhaps the most compelling rationale for employing no or a low discount rate is what is known as intergenerational equity. We may not fully appreciate the value of protecting natural resources off into the future. Thus, it would be reckless to waste resources today that may later be found to have great value (utilitarian or otherwise) to future generations.[34] Thus, we should be prudent in how much we discount the environmental benefits that natural systems hold for future generations. Discounting the future value of natural resources and ecological services is antithetical to the concept of valuing our environmental legacy.

When discounting takes place over a long time frame, such as the legacy period, the protection of resources for the future appears to have little or no value today. Consequently, to further the Act's goal of ensuring an environmental legacy, a different approach is needed. Emergy accounting does not discount the value of protecting resources for the future. Accordingly, emergy accounting can be a useful tool in evaluating legacy impacts under the Legacy Act. Emergy accounting not only rejects discounting but also recognizes that, as resources become rarer, their value becomes greater.

There are two cases where scarcity increases emergy value. First, from a donor perspective, in environments where scarcity necessitates more input to make something, the emergy required for the same final product is greater, and thus its value is greater. A good example is the difference in transformity of rainfall in an area of abundant moisture (e.g., tropical rainforest) and a region of scarce moisture (e.g., desert). The desert requires nearly fifty times the sunlight investment to generate the same rainfall as that in the rainforest; thus, the emergy value of desert rain is fifty times that of rainforest rain.

The second case is related to the effect of scarcity on reproducibility. As resources, especially living organisms, become scarcer the ability to reproduce them becomes more and more difficult. The time and energy required to replace the last individual of a species is tremendous when one realizes that to do so requires thousands of years of evolution to re-evolve it from its nearest relative.

Strengths and Weaknesses of the Method

The greatest strength of using emergy as a measure of value is that all things are put on a common scale that is objective and scientific. Products and processes in the economy of humanity and nature are valued according to what it took to make them rather than what people think they are worth. This results in the ability to compare the effects of alternatives across broad and diverse categories of things. Because the economic and ecological consequences of social value choices are expressed on a common scale, emergy synthesis eliminates the uncertainty associated with the controversy over how to quantify and compare the different values.

Perhaps the greatest problem in using emergy synthesis is that it is a complex method that requires broad knowledge and special training by its practitioners. Emergy synthesis is also a fairly new methodology that at first received fierce opposition in the economics community. As the true relationship between our economy and its support environment has become more apparent, economists have become more open to new ways of thinking, as illustrated by the development of a new field, ecological economics, and the increased interest in alternative valuing systems.

A significant challenge to gaining acceptance of emergy synthesis is the fact that it is a donor system of value. There are two general methods by which value can be determined: (1) receiver-based systems (e.g., the market economy based on human willingness to pay) and (2) donor-based systems (based on the production cost of goods and services).[35] Donor-based methods of determining value are not usually employed to determine value in Western societies, where value is invariably associated with the subjective idea of utility to humans. Because Western societies are more inclined to measure value from a receiver-based perspective, it has been difficult to alter this mind-set for environmental management decisions that require a broader, integrative viewpoint. A main impediment to the use of emergy synthesis in decision making is society's reluctance to change value systems under the perceived notion that willingness to pay works in the marketplace and therefore can be made to work in the public policy arena.

Values determined by a donor method are based on objective facts and scientific principles. Both objective and subjective values can be determined using statistical methods with known limits of confidence, and both measures are dependent on the accuracy of the information used in the determination. Donor-based estimates of value are limited only by the available information, and these limits may be less confining than in the receiver-based method, as receiver-based estimates of value are limited by the information available to the general public, when it is canvassed to obtain value estimates.

Conclusion

The National Environmental Legacy Act provides a new approach to protecting our natural resources for future generations by ensuring that our natural resources are used in a sustainable manner. Implementation of the Act would require a methodology that enables us to quantitatively and qualitatively evaluate whether resources are being affected in such a way as to degrade or deplete our natural legacy. Emergy synthesis is a well-developed methodology that can be employed to fill this need. Emergy synthesis has many benefits over other methodologies. Most significantly, emergy synthesis does not rely on consumer preferences to determine the value of a resource. Instead, it uses a donor method of valuation that more accurately assess the intrinsic value of the resources. Moreover, emergy synthesis captures the quality of the value of the resource in addition to the quantity. Finally, by explicitly rejecting the concept of discounting, emergy synthesis inherently provides a long-term, or legacy, view of resource utilization. Not only can emergy synthesis be used to evaluate the quantity and quality of resource degradation or depletion over the legacy period, but it also can be employed in the enforcement, monitoring, and exceptions components of the Legacy Act.

NOTES

1. Emergy is a contraction of "energy memory." Similar to embodied energy, emergy is the sum of all the inputs required to make resources, goods, and services, expressed in units of energy of the same form. H.T. ODUM, ENVIRONMENTAL ACCOUNTING: EMERGY AND ENVIRONMENTAL DECISION MAKING (1995).
2. For a detailed discussion of the origins of the emergy concept, *see* M.T. Brown & S. Ulgiati, *Energy Quality, Emergy, and Transformity: H.T. Odum's Contributions to Quantifying and Understanding Systems*, 78(1–2) ECOLOGICAL MODELING 201–13 (2004).
3. E.P. ODUM (WITH H.T. ODUM), FUNDAMENTALS OF ECOLOGY (1953).
4. "Beginning in the last century man began to develop an entirely new basis for power with the use of coal, oil, and other stored-energy sources to supplement solar energy. Concentrated inputs of power whose accumulation had been the work of billions of acres of solar energy, became available for manipulation by man." H.T. ODUM, ENVIRONMENT POWER AND SOCIETY (1971).
5. One of the first publications of Odum's concept of energy quality was in the acceptance speech for the Swedish Crafoord Prize, awarded by the Royal Swedish Academy of Sciences. The acceptance speech was later published in *Ambio*. H.T. Odum, *Energy, Ecology and Economics: Royal Swedish Academy of Science* 2(6) AMBIO 220–27 (1973).
6. The term "emergy" was coined by David Scienceman, from Australia, who was a visiting scientist at the University of Florida working with Odum's research group during the late 1980s. D. Scienceman, *Energy and Emergy*, in ENVIRONMENTAL ECONOMICS: THE ANALYSIS OF A MAJOR INTERFACE 257–76 (G. Pillet & T. Murota eds. 1987).
7. Brown & Ulgiati, *supra* note 2.
8. M.T. Brown & S. Ulgiati, *Emergy Evaluation of Natural Capital and Biosphere Services*, 28(6) AMBIO 1 (1999).
9. C.P. STICKNEY & R. L. WEIL, FINANCIAL ACCOUNTING: AN INTRODUCTION TO CONCEPTS, METHODS, AND USES (10th ed. 2003).
10. D. Campbell, *Financial Accounting Methods to Further Develop and Communicate Environmental Accounting Using Emergy*, in EMERGY SYNTHESIS 3: THEORY AND APPLICATION OF THE EMERGY METHODOLOGY – PROCEEDINGS OF THE 3RD BIENNIAL EMERGY CONFERENCE 185–198 (M.T. Brown ed. 2005).
11. Odum, *supra* note 1.
12. The methodology is developed in great detail in Odum, *supra* note 1.
13. As used here, the term "metric" means a standard for measuring or evaluating something, a basis for assessment.
14. Gaylord Nelson, *The Bankruptcy Files*, WILDERNESS Vol. 57, Issue 205 (Summer 1994), 10.
15. Campbell, *supra* note 10.
16. Resource quality to some means the concentration of minerals or other natural resources in sufficient quantities to make them economically extractable.

See, e.g., Cutler J. Cleveland et al., *Natural Resource Quality*, in ENCYCLOPEDIA OF EARTH (Robert Costanza & Cutler J. Cleveland eds. 2007), *available at* http://www.eoearth.org/article/Natural_resource_quality.

17. Brown & Ulgiati *supra* note 2, at 201–13.
18. These studies are summarized in Table 4.1. Fields of study, emergy projects, and references appear in Brown and Ulgiati, *supra* note 2.
19. M.T. BROWN & E. CAMPBELL, EVALUATION OF NATURAL CAPITAL AND ENVIRONMENTAL SERVICES OF U.S. NATIONAL FORESTS USING EMERGY SYNTHESIS: FINAL CONTRACT REPORT TO USDA FOREST SERVICE UNDER COOPERATIVE AGREEMENT NUMBER 05-DG-11120101–019 (2007).
20. *See, e.g.*, SIDNEY A. SHAPIRO & ROBERT L. GLICKSMAN, RISK REGULATION AT RISK: RESTORING A PRAGMATIC APPROACH (2003); DANIEL A. FARBER, ECO-PRAGMATISM: MAKING SENSIBLE ENVIRONMENTAL DECISIONS IN AN UNCERTAIN WORLD (1999).
21. Because there are no significant markets for most environmental services, cost-benefit analyses, preparation of environmental-impact statements, wetlands mitigation banking, Superfund remediations, and oil-spill cleanups often ignore these services. James Salzman, Barton Thompson & Gretchen Dailey, *Protecting Environmental Services: Science, Economics, and Law*, 20 STAN. ENVTL. L.J. 309, 311 (2001).
22. *See, e.g.*, John M. Heyde, *Is Contingent Valuation Worth the Trouble?* 62 U. CHI. L. REV. 331 (1995); FARBER, *supra* note 20, at 47–51, 84–87, and 99–101.
23. FARBER, *supra* note 20, at 99–101.
24. *Id.* at 100.
25. Brown & Ulgiati, *supra* note 8, at 9.
26. *See, e.g.,* HERMAN E. DALY & JOSHUA FARLEY, ECOLOGICAL ECONOMICS: PRINCIPLES AND APPLICATIONS (Island Press 2004). Salzman et al., *supra* note 21; G.C. Daily et al., *The Value of Nature and the Nature of Value*, 289 SCIENCE 395–96 (2000); J.B. Ruhl & James Salzman, *The Law and Policy Beginnings of Ecosystem Services*, 22 J. LAND USE & ENVTL. L. 2 (2007).
27. Brown & Ulgiati, *supra* note 8, at 1.
28. Jorge L. Hau & Bhavik R. Bakshi, Promise and Problems of Emergy Analysis, 178 ECOLOGICAL MODELING 215, 216 (2004). Although criticisms have been leveled at emergy synthesis, they are primarily based on a lack of understanding on the part of the critics, insufficient communication of emergy theory outside of the scientific world of emergy scholars, lack of clear links with related concepts in other disciplines, and the types of general criticisms that are often directed at new, groundbreaking ideas. *Id.* at 223 (reviewing criticisms of emergy and concluding that many of the criticisms leveled apply not just to emergy analysis but to all methods that focus on a holistic view). A list of publications that provide critiques of emergy analysis includes R.U. AYRES, ECOLOGY VS. ECONOMICS: CONFUSING PRODUCTION AND CONSUMPTION, CENTER OF THE MANAGEMENT OF ENVIRONMENTAL RESOURCES (INSEAD 1998); C.J. Cleveland, R.K. Kaufmann & D.I. Stern, *Aggregation and the Role of Energy in the Economy*, 32 ECOLOGICAL ECON. 301–17 (2000);

B.A. Mansson & J.M. McGlade, *Ecology, Thermodynamics and H.T. Odum's Conjectures*, 93 OECOLOGIA 582–96 (1993); D.T. Spreng, NET-ENERGY ANALYSIS AND THE ENERGY REQUIREMENTS OF ENERGY SYSTEMS 289 (1988).

29. M.T. Brown & A. Buenfil, *Emergy Synthesis as a Tool for Evaluating Management Options for Fresh Water in Africa*, in AQUATIC CONSERVATION AND MANAGEMENT IN AFRICA (T.L. Crisman, L.J. Chapman, C.A. Chapman & L. Kaufman eds. 2002).
30. M.T. Brown & S. Ulgiati, *Emergy Measures of Carrying Capacity to Evaluate Economic Investments,* 22(5) POPULATION AND ENVIRONMENT 471–501 (2001).
31. M.T. Brown & T. McClanahan, *Emergy Analysis Perspectives for Thailand and Mekong River Dam Proposal,* 91 ECOLOGICAL MODELING 105–30 (1996).
32. Farber, *supra* note 20, at 133.
33. J.B. Ruhl, *Working Both (Positivist) Ends Toward a New (Pragmatic) Middle in Environmental Law*, 68 GEO. WASH. L. REV. 522, 539 (2000) (book review).
34. Farber, *supra* note 20, at 151–52.
35. Odum, *supra* note 1.

5 Citizen Science and the Next Generation of Environmental Law

Christine Overdevest and Brian Mayer

IF THE NATIONAL ENVIRONMENTAL LEGACY ACT IS TO SET ACCURATE baseline measures of the quantity and quality of our natural resources, whose information should be included in the establishment and monitoring of these environmental metrics? This chapter argues that the Legacy Act should pay attention to where the law might cultivate a more active citizenry by enlisting citizen volunteers in the collection and diffusion of environmental data and the public education about environmental quality. One approach to cultivating such an active citizenry is through public participation in various forms of citizen science. By "citizen science," we mean projects involving citizen volunteers in the collection, analysis, and interpretation of data through which the public becomes more engaged in federal research and decision making. Citizen science programs can increase the amount of information about the ecological condition of local resources, support data sharing for community education, build networks, leverage local resources, and help provide information to motivate local civic conservation action. In turn, participation in the collection and discussion of environmental data creates opportunities for the publicity necessary to cultivate a more active and informed citizenry. In addition, citizen science can create informal social pressure for regulatory compliance and provide support for agency and citizen enforcement efforts.

To develop the benefits of active participation and citizen science, this chapter draws on the political philosophy of Hannah Arendt. Hannah Arendt focused on the importance of active citizenship in society, arguing active involvement yields citizens with the capacity for effective political participation. To illustrate the potential of citizen science in establishing active citizenship, the authors examine the case of citizen monitoring of local air quality using lay-generated technology called "bucket brigades". This case illustrates how environmental law can draw creatively on citizens as resources for monitoring and enforcement at the same time that the law can create opportunities for more active public participation.

Active versus Passive Citizenship and Environmental Law

The first generation of federal environmental law gave rise to what is commonly thought of today as the participatory paradigm of engaging the public in agency environmental planning and decision making. With the exception of the citizen suit provision, however, which has encouraged active private enforcement of environmental statutes (May 2003), the first generation envisioned a fairly limited and passive role for interested and affected citizens. Agencies were required to publicize proposed environmental rules and create periods for open public comment. Administrators weighed and balanced the input from the public against the agency's mission and available scientific information. Critics have pointed out that agency personnel often assume that nonexpert publics are underinformed, especially when public views diverge from those of scientific or agency experts. Consequently, agency officials often consider public input less significant than nonlay forms of expertise (Eden 1996; Kleinman 2000). Similarly, others have argued that the open-ended balancing tests of federal environmental statutes and the broad discretion of administrators do little to mitigate agency biases (Flournoy, Chapter 1 in this volume). Furthermore, participants in traditional notice-and-comment processes do not benefit from the capacity building and learning that the more active participation associated with an engaged citizenry can produce.

The second generation of environmental law, with its emphasis on increasing the efficiency of environmental policy by means of market mechanisms, has done little to address the democratic lacunae. If anything, market-based, cost-benefit analysis encourages greater reliance on technocratic decision processes, thus creating more distance between policy legitimation and citizen participation and learning (Ackerman and Heinzerling 2004). Market-based policies remove citizens from processes of civic reflection and undermine the firsthand experience and learning that underpin effective citizenship.

The twentieth-century political philosopher and public intellectual Hannah Arendt provides a compelling critique of such passive forms of participation and promotes the benefits of active citizenship. In Arendt's view, public participation properly understood is not merely a means of weighing public support for policies so much as it is a way of developing a citizenry with the capacity for critical political reflection.

The Alternative of Active Citizenship

Since Alexis de Tocqueville's (2000) famed trip to see democracy in America firsthand, he and subsequent observers have feared that the scale of modern society might undermine the United States' experiment in democratic participation. Hannah Arendt, writing in the 1950s, shared Tocqueville's worries that, in a large-scale democracy, citizens would retreat into a culture of individualism, pursue the satisfaction of private needs, and undercut the legitimacy of their own democracies. In the United States today, voter turnout suggests reason for concern. The United States has among the lowest voter turnout rates in the world. In the 2006 statewide midterm elections, turnout was an all-time low of 15.4 percent of eligible voters, according to the Center for the Study of the American Electorate (CSAE 2006). In the 2008 presidential primary, twenty-three of thirty-four states holding primaries had record turnouts, but the overall turnout was less than one-third of eligible voters (30.2 percent) (CSAE 2008). Today's social critics lament that most Americans "bowl alone" (Putnam 2000); that is, they do not participate in shared activities in ways that might engender broader concerns about and active engagement in their communities. Public advocacy is typically performed by professionalized advocacy groups, which purport to represent the public while actually having very little direct contact with average citizens (Skocpol 2003). Many observers fear that this trend demonstrates that the American public is an apathetic lot and that, consequently, the quality of our democracy suffers.

For the next generation of environmental law to succeed, this challenge to democratic participation must be addressed. Hannah Arendt wrote extensively about why active participation in public affairs is important. Active citizenship involves engaging in activities affecting one's broader community and, according to Arendt, enlarges mentalities and helps counteract apathy (Arendt 1958). She argued that civic engagement in collective rather than individual social problems helps citizens learn about public problems and share knowledge of those problems with one another. Hearing others' points of view allows individuals to assess and evaluate their own assumptions and biases. Through the process of publicly testing one's own reasoning against the reasoning of others, citizens develop the capacity for critical reflection and more informed political debate. In this way, Arendt believed that citizens' capacities as effective political agents developed in the process of shared political debate and activity. If critical thinking forms in the process, then public participation is not merely a means of creating more legitimate policies so much as a way to develop a citizenry with greater capacity for effective citizenship.

It is worth noting that, for Arendt, the audience for citizen participation was as important as the participation itself (Arendt 1958). Without a broad audience of citizens from the community, the prospects of expanding one's views, assessing the reasonability of assumptions, and achieving public learning is greatly limited. It is the reaction of an audience that helps to judge the reasonability of action.

From this perspective, the problem with the first generation of environmental law is that the audience for public input often tends to be limited to administrators. Those administrators do not provide a proper audience for citizens because their interactions with citizens are highly formalized, one sided, and indirect. Administrators often make decisions removed from citizens' discussions, where political opinions are formed. Citizens cannot serve as an audience to regulators because administrators collect input and weigh and balance it away from public view. Thus, often there has been no proper audience in the first generation of environmental law.

For this reason, the Arendtian view of environmental law calls not only for greater active participation but also for a broader audience to judge the reasonability of input. The roles of citizens as witnesses and potential narrators (i.e., an active citizenry broadly construed) are critical to cultivating an effective public sphere. Arendt sometimes referred to this marriage of activity, speech and/or disclosure, and audience as publicity (Arendt 1982: 41). Drawing on the work of Immanuel Kant, she argues: "we may safely say that the external power which deprives man of the freedom to communicate *his thoughts publicly also takes away his freedom to think*, the only treasure left to us in our civic life and through which alone there may be a remedy against all evils of the present state of affairs" (Arendt 1982: 41). Arendt continues: "it is precisely by applying [others'] critical standards to one's own thought that one learns the art of critical thought. And the application one cannot learn without publicity, without the testing that arises from contact with other people's thinking" (Arendt 1982: 42). Although Arendt argued for its importance, she recognized that an active community is not easy to construct. She was fully aware that people could easily conduct their whole lives without any form of political participation. Creation of an active community was a cultural achievement that required collective resources to establish. Even the ancient Greeks, Arendt points out, realized this – first experimenting with fines for absences in public dialogues and then paying citizens to attend assembly meetings (St. Croix 1975).

Arendt also recognized that, depending on the extent of participation, publicity happened in degrees. A political discourse marked by exchange of experience and opinion over coffee among neighbors marked one degree of publicity, whereas a much greater degree of publicity could be achieved in a larger forum. The more inclusive the gathering, the greater the publicity.

From this point of view, the role of participation in the third generation of environmental law is to reconnect citizens to active participation. We argue for the Legacy Act to break fundamentally with the passive, formalistic, one-sided participation models of prior U.S. environmental law. One potential way to do so is to address the goal of better tracking of environmental baselines and deviations from baselines by encouraging citizen science. Citizen science is an active form of public participation that can also increase the audience, shared discourse, and publicity about local environmental problems.

Active Participation and the Legacy Act

The Legacy Act argues for defining in concrete terms the environmental legacy we wish to leave future generations, requires federal collection of baseline information and ongoing monitoring of natural resource quality, and shifts the burden of scientific justification to parties who wish to deplete or degrade resources. This solution to intergenerational environmental inequity envisions a more precautionary approach to achieve the goals of protecting a national resource legacy. To help monitor environmental quality, the Act proposes to set benchmarks and metrics drawing on independent scientific panels to define them and tracking quality measures to help us not fall victim to the problem of shifting baselines, as discussed by

Ankersen and Regan (Chapter 3, in this volume). Still, the Legacy Act recognizes that an open-ended demand for information could prove an excessively costly quest.

The next generation of environmental law can realize a more cost-effective, richer base of information about environmental quality by creatively drawing on resources from civil society at the same time that it encourages a more active citizenry. Environmental law can cultivate a more active citizenry by enlisting citizen volunteers, sometimes called "citizen scientists," in environmental data collection, diffusion, and public education about environmental quality. Citizen science programs increase the amount of information about the ecological condition of local resources, support data sharing for community education, and build networks of citizens who are conversant about local environmental quality. Information collected by citizen volunteers can help build social networks of citizens knowledgeable about changing local environmental quality and can thereby increase public discussion about environmental quality. Information collected by citizen volunteers also can be used as a source of informal social pressure for compliance and can provide support for agency and citizen enforcement efforts. In this way, citizen science can be an important institutional design for achieving a more active and informed citizenry in the next generation of environmental law.

The Legacy Act should therefore create a role for citizen science groups to collect data on key metrics and share it locally. To illustrate the potential of citizen science, we describe the extent of its current use in the United States and then focus on bucket brigades as an illustrative case of active participation that creatively draws on civil society.

Citizen Science: Creatively Empowering Civil Society

A great deal of interest in citizen science has arisen over the past twenty years. In fact, citizen volunteer monitoring programs have become widely popular across North America (Environmental Protection Agency [EPA] 1998). Citizen-scientists sometimes work with researchers as part of scientific research teams, but groups also develop from the bottom up – collecting data to address local conservation projects, environmental education, or advocacy goals. Many volunteer monitoring groups are organized by state and university cooperative education programs. Thirty states have such cooperative programs. Other groups are organized by national and regional conservation and environmental nongovernmental organizations (NGOs), like the 110-year-old Audubon Christmas Bird Count. The *National Directory of Volunteer Environmental Monitoring Programs,* a national survey last conducted in 1998 identified 772 environmental monitoring organizations and made a conservative estimate of more than a half million volunteers (EPA 1998). Between 1994 and 1998, the number of groups grew by 67 percent. The annual median budget for groups was $2,000, although 44 percent operated on a budget of $1,000 or less. Eighty-four percent reported using their data for education; 67 percent, for establishing baselines; 61 percent, to screen for problems; 53 percent, for research; 46 percent, for community organizing; 43 percent, for advocacy; and 43 percent, for local watershed planning. Groups monitor lakes, streams, beaches, wetlands, groundwater, marine environments, air, and watersheds. Volunteers monitor physical, biological, and chemical parameters such as stream flow rates, invasive species, and bird abundance, and undertake tests for bacteria and toxic chemicals in our air and water.

The Power and Limitations of Citizen Science: The Case of Bucket Brigades

Despite the existence of both federal and state environmental laws regulating industrial emissions of hazardous pollution in the United States, industrial and other polluters in the United States release enough hazardous pollution into the air to place more than 92 percent of the U.S. population at an increased risk of developing respiratory disease (EPA 1999). Current federal environmental law currently requires only limited monitoring of air quality, which provides only rough estimates of air quality over large regions of the United States (McGarity 2008; see also EPA 2008). This level of data resolution does not inform regulators or the public about local pollution sources that may put public health at risk. In response, citizen science groups in an estimated one hundred communities have organized bucket brigades to collect site-specific air-monitoring information to better characterize local air quality and to identify local sources of pollution.

Bucket brigades first emerged to gather information for a civil lawsuit following an industrial accident at an oil refinery in Rodeo, California (for a more detailed history of their emergence, see Overdevest and Mayer 2008).

A San Francisco community environmental organization subsequently formalized the monitoring model, providing training to grassroots citizens' groups on how to develop their own bucket brigades in communities with complaints about local industrial odors, leaks, or other indicators of toxic releases. Bucket brigades allow citizen groups to conduct their own empirical investigation of local air quality. In bucket brigades, community members assume one of several roles. Sniffers are residents who are trained to record the date, time, and place of any chemical smells, odors, smoke, or unusual vapor clouds coming over the fence lines of local plants. When sniffers observe evidence of potential toxic emissions, typically from EPA-permitted factories, they call samplers – local citizens trained in the use of air-quality sampling devices. Samplers respond by taking samples and sending them to a certified air-quality lab for chemical analysis. Brigades use the results to inform neighbors of the levels of pollutants in neighborhood air; send press releases to local newspapers; and share their findings with firms, government officials, and at-risk populations.

Bucket brigades increase the amount of data available about the ecological condition of local public resources. When bucket brigades make alarming discoveries about departures from federally mandated air-quality standards, they present the data to regulators and firms and demand tougher enforcement and compliance. They use their data as an important component of their public narratives about local environmental quality. They share their data-rich stories with the local and state newspapers and other media in an attempt to inform the broader public of local air-quality problems. They create additional publicity through their attempts to involve universities, public advocacy groups, and other NGOs in their clean-air campaigns. In this way, they help broaden the witnesses, narrators, and audiences that Arendt viewed as instrumental in an effective public sphere. In the following discussion, we draw from accounts of bucket brigades, including interviews with brigade members, newspaper stories, and other published sources, to illustrate how creatively drawing on civil-society resources can achieve public goals of building a more active and informed citizenry.

The Power of Citizen Data

There are various ways that citizen monitoring create publicity and broaden the audience for reflection and critical attitude development. First, citizen monitoring encourages active participation of volunteer citizens in community problems and encourages learning about local resource issues. By frequently participating in measurement activities, volunteers learn first-hand about levels of local air quality and changes in quality, and they gain insight into polluters and their impacts on public resources. The increased learning about the specific challenges and problems facing the local community can serve to motivate citizens to discuss their concerns with others, including directly approaching firms and regulators to insist that state and federal standards be maintained (Overdevest, Orr, and Stepenuck 2004). Thus, citizen monitoring increases the audience for local environmental problems by motivating citizen volunteers to follow up on their concerns in the local community, thereby putting firms and regulators in a position to provide a public account of their behavior. The Legacy Act can benefit from the additional watchdog or citizen enforcement efforts this may help to create.

Several examples from buckets brigades illustrate how monitoring data motivates citizen action and translates citizens' generalized, nonspecific grievances (e.g., complaints about a foul odor) into a specific complaint, often against a particular polluter – a type of complaint that regulators or the firm must explain and defend. For instance, Anne Rolfes, director of the Louisiana Bucket Brigade, a support organization that facilitates several community bucket brigades in the industrial district known as Cancer Alley, reports that, since their initiation in 2000, bucket brigades have resulted in several actions, including an EPA notice of violation, an EPA investigation into Louisiana's environmental programs, and a voluntary buyout of three hundred families by a refinery complex. Rolfes, reflecting on how she believes bucket brigades work, recalled, "It changes a very vague complaint, like 'It smelled bad last week,' to 'You violated the state benzene standard.' That's the beauty of it" (Strickland 2004). In this case, the sampler, a member of the local community, had approached the firm to request an explanation or account of why there was a departure from state benzene standards in the air around the facility. In this way, monitoring data translated citizens' generalized, nonspecific grievances into a specific complaint against a particular polluter. Citizens' monitoring data appears to be an important resource in motivating citizens to approach firms and ask for a public account about departures from specific regulatory standards.

In another case, on December 8, 1998, an oil refinery in Norco, Louisiana, had an upset, defined as an unplanned, unpermitted release. The company publicly claimed that no hazardous chemicals were released in the upset, but the local active bucket brigades had captured samples that indicated otherwise. Iris Carter, a local citizen of Norco, Louisiana, and member of a bucket brigade trying to hold the refinery to public account, stated, "When we had scientific evidence, we could say, 'You did release this stuff' " (Nijhuis 2003). The director of the Louisiana brigade organization explained how bucket brigades motivated citizen action instead of citizen apathy: "People had something to do. In the middle of a [release] event, they could take action" (Nijhuis 2003). In this case, after several years of citizen monitoring and other forms of activism, the refinery agreed to a voluntary property purchase program, providing an $80,000 allowance for three hundred families closest to the fence line of the chemical and refinery complex to relocate. As an EPA air-quality analyst commented when asked about the role of citizen monitors as regulatory watchdogs, "The EPA can't be in everybody's backyard at the same time.... It's always useful to have other eyes out there" (Nijhuis 2003).

Citizen-scientists use their data in more ways than simply complaining directly to polluting firms or regulators. The data become an important component of narratives about local environmental quality that are shared more broadly in the local community. The data inspire discussion and publicity about local environmental quality trends. Constructing public narratives can increase the number of local individuals conversant and engaged with local public problems related to environmental quality and departures from quality. The bucket-brigade case illustrates this with respect to a number of different actor networks, including extensive public diffusion of the information through the media, and attempts to network with specific publics such as scientific research professionals, NGOs, and advocacy groups, as well as the regulatory community. Thus, citizen science enlarges the audience well beyond the citizen volunteers themselves.

For example, citizens' monitoring data become a component of public narratives when data is shared with local media. The media is an important resource for diffusing information and setting agendas (McCombs and Shaw 1972). The media help share the claims of regulatory failure and firm intransigence with a broader public and, in the process, helps creates an enlarged audience to which public accounts are due. Furthermore, discussion in the general media about a small group of concerned citizens may work to generate wider interest in environmental monitoring campaigns and involvement in environmental management. Through the documentation of successes and failures in achieving state and federal environmental performance goals in various media outlets, bucket brigades and other similar forms of citizen science set public agendas and encourage action on environmental quality. In the same way, active citizens monitoring for compliance with the Legacy Act could enlist the media to generate wider interest in monitoring and protecting our natural resource legacy.

For example, bucket-brigade activists, after taking samples in which regulatory limits have been surpassed, routinely send letters detailing the outcomes of the sampling to regulators and firms, and send copies as press releases to major media outlets. Bucket brigades have a specific strategy of wide public diffusion that appears to be successful in part because the media views citizen-collected information about local departures from regulatory standards as newsworthy.

The Louisiana Bucket Brigade (2009) reports that, after a concerted effort, more than sixty media stories covered the Concerned Citizens of Norco campaign. These letters and press releases describe regulatory requirements and community sampling efforts and identify departures. Typically, they identify where in the community sampling took place and report the results; identify the known and possible health effects of exposures; identify especially vulnerable populations, such as children's schools, day-care centers, and homes for the elderly; and appeal to regulators and to the firm for relief. Bucket-brigade information has appeared in numerous mainstream news publications, including the *Miami Herald*, the *Houston Chronicle*, the *San Francisco Chronicle*, the *Philadelphia Inquirer*, *USA Today*, and the *Christian Science Monitor*, as well as such outlets as *Mother Jones* and *Grist*, and in many more local papers in towns where facilities operate.

The bucket brigade in the town of Tonawanda in western New York, which sampled near two local plants, illustrates the use of a press release to achieve these ends. As the organizer, Tim Logsdon, director of Clean Air Coalition, explained, the point of the sampling is to educate the broader public about exposure: "We took these samples because the people living here have the right to know what is in the air that they are breathing, and what risks they may be taking by breathing these chemicals" (Sapong 2005: 1). The news article written about the Tonawanda sampling is largely reflective of concerns raised by citizen-scientists. It points out that findings indicated that levels of carbon disulfide found were four hundred times higher than allowed by the state. The levels of benzene were higher than allowed in Texas, Louisiana, and North Carolina. Finally, the article reports that the bucket brigade requested that the state Department of Environmental Quality, the EPA, and the state attorney general's office investigate and "create a plan to reduce the pollutants.... The groups also want community and environmental organizations to be involved in an effort to create a comprehensive air-monitoring procedure for the area."

Bucket brigades use their information and their accounts of local pollution to create alliances and build networks of concerned citizens in the community. The broader array of citizens and citizen groups help build a broader base of support for clean air. We illustrate some examples of the power of citizen monitoring to create alliances and create more active community networks. For instance, bucket brigades often create relationships with local, state, and regional NGOs and advocacy groups to pursue their claims against polluters. An active Philadelphia bucket brigade, the Community Labor Refinery Tracking Committee (CLRTC), joined with the Environmental Integrity Project, a Washington, D.C.-based nonprofit organization, to document claims that a local Sunoco refinery complex had committed violations of federal law, the Pennsylvania Code, and Sunoco's permits. The Mid-Atlantic Environmental Law Group represented the bucket brigade in a lawsuit, filed in April 2005. In November of that year, Sunoco settled the suit, agreeing to install new pollution-reduction technologies and purchasing sophisticated real-time air-monitoring equipment for the brigade. These expensive real-time monitors, which provide on-the-spot data analysis and ongoing monitoring, allow the Philadelphia bucket brigade to obtain immediate readings of pollution, saving them the expense of sending samples to a laboratory.

Wildlaw, a nonprofit environmental law firm based in the southern United States, has helped residents of Ocala, Florida, implement bucket-brigade monitoring of the Royal Oak Charcoal plant. The Tulane University Environmental Law Clinic joined a lawsuit by the St. Bernard Citizens for Environmental Quality after bucket brigades showed violation of benzene standards by a firm in St. Bernard Parish.

University scientists have also become involved in bucket-brigade campaigns. The University of Texas toxicology professor Marvin Legator completed health studies in Port Arthur after the Community In-power and Development Association (CIDA) founder Hilton Kelley led a bucket-brigade campaign there. Other academic institutions, such as York University, the University of Illinois' School of Public Health, the University of Pennsylvania, and the University of Michigan's School of Public Health, have aided or participated in the sampling and analyses of air toxics using bucket brigades. In this way, citizen science campaigns mobilize additional resources around the issue of environmental quality that can benefit the Legacy Act and help meet a vision of a more active citizenry.

Deconstructing Citizen Science

Although there are benefits of generating local citizen monitoring and data-sharing campaigns, there are also several weaknesses and limitations to the citizen science model that need to be addressed and of which Legacy Act should take account. Even with the sampling protocols bucket brigades follow, a reasonable objection to community-generated data is that the data are collected in uncontrolled situations by nonexperts who may intentionally or unintentionally misrepresent the data. Moreover, several polluters are often located in close proximity in industrial districts, such as Cancer Alley, where the Louisiana Bucket Brigade operates, or Refinery Row, in southwestern Philadelphia, where the Community Labor Refinery Tracking Committee collects samples. This makes accurate attribution of pollution sources methodologically challenging. Studies of a variety of citizen science organizations have shown that critics of lay forms of science often attack the methods used rather than directly disputing the findings of citizens (Brown 2007). For instance, in response to the sampling of the Calumet bucket brigade on Chicago's East Side that showed five refinery chemicals registering above the state's "level of concern," a British Petroleum spokesperson retorted: "[a]s for individuals or groups taking air samples, we are not aware of the duration, quality, location, or method of their sampling. We encourage them to conduct their activities with the full knowledge and oversight of [the Illinois Department of Environmental Management], just as we do." As another methodological critique, the British Petroleum spokesperson pointed out that as many as fifty other companies in the area of the street sampled by activists also release toxics (Laasby 2007).

Although industry critics neglect the fact that a lack of citizen-collected data often means no data at all, it is still important for the Legacy Act to address methodological concerns about citizen-collected data. There are a variety of ways the Legacy Act could encourage more local monitoring while improving the overall quality of citizen monitoring. Given the importance of the quality of citizen-collected data, we examine the potential role of the Legacy Act in the following section.

Government and Citizen Monitoring: How the Legacy Act Provides an Opportunity to Enhance Accountability of Citizen Monitoring

The Legacy Act can help civil society regulatory groups avoid potential methodological pitfalls and related challenges from affected actors through education and capacity development. Developing data-sampling protocols and guidelines, encouraging training and certification in citizen sampling, and providing quality-assurance tests for citizen-sampling devices are all ways that the Legacy Act can facilitate the robustness of citizen monitoring. This would require that the involvement of citizen-scientists potentially be considered by the proposed Committee of Experts, as the committee members work with agencies to develop metrics in order to ensure that some metrics will encourage and permit citizen monitoring.

Currently, the EPA requires that any monitoring data it uses be "scientifically valid, defensible, and of known precision and accuracy," including data that is "supported or mandated through contracts, regulation or other formalized agreement" (EPA 1979: 1–2). The Legacy Act might be drafted to extend the ability of citizen groups to enter into agreements pursuant to which stewardship agencies would undertake quality assurance on technologies such as bucket brigades. Stewardship agencies could play an important role in addition to their traditional enforcement roles by helping to monitor the legitimacy of citizens' monitoring technologies.

Resources, Training, and Rewards

The operation of community environmental monitoring requires an initial investment in monitoring technology. Despite the relatively low cost of lay data collection, the expense of equipment – and for some forms of monitoring, the expenditures for analysis of the samples – may be too high for some smaller community organizations. This suggests a second important way for environmental regulatory policy, and the Legacy Act in particular, to support citizen science. First, agencies under the Legacy Act could be encouraged to spread their funds for baseline and ongoing data collection among in-house and civil society monitoring organizations. Another option is to build on the current support for community volunteer monitoring efforts already organized at the state level. For instance, today, thirty state and territory cooperative extension offices have sponsored volunteer monitoring programs. States create these programs largely for public education and outreach reasons. However, the funding already allocated could be used to achieve the goals of expanded federal monitoring as well.

Relevant agencies could offer training for citizens. By building a community of local volunteers in communities across the nation, the government can draw on resources of citizen volunteers for compliance at the same time that it creates meaningful education and active participation opportunities. Broader social benefits from citizen participation in these activities include public learning, increased information about local environmental parameters, and closer interaction between the public and the regulatory community.

Communities in which citizen reporting leads to federal fines or penalties could be rewarded with bounties or rewards under the Legacy Act. Perhaps some portion of fines could be awarded to citizens' monitoring organizations to help sustain their activities.

Defining Permissible Uses of Citizen-Collected Data and Setting Data Standards

Even with efforts such as these to address its weaknesses, citizen science is still likely to be subject to methodological and related critiques. This raises the questions as to which purposes citizen-collected data should be used for and the standards citizen-collected data should meet. Should it be used directly to justify agency enforcement action? Should citizen data be used only informally by citizens? If it meets appropriate regulatory standards, should it be relied on by agencies developing plans under the Legacy Act?

An illustration from the bucket-brigade case suggests one resolution of these questions. Following a number of industrial accidents at several different facilities that significantly affected public trust in regulators, the EPA funded a community bucket brigade in Contra Costa County, California. The Contra Costa County Health Services Department trained samplers and sniffers using funds from the EPA. A quality-assurance plan was required because of the usage of EPA funds. Field tests compared bucket sampling to standard Summa canisters to ensure that volunteers were using the samplers correctly and that the equipment met requisite technical criteria for accuracy. For the most part, the samples taken with the buckets were within the acceptable range of those from the more official Summa canisters (Hobson and Fishman 1998). Following analysis, the quality-assurance report released by the EPA set out public education as the official purpose that the data were intended to serve, but it also suggested ways that information could be used more broadly as a source of informal social control and as a resource for generating greater publicity. This is exactly the marriage of activity, speech, disclosure, and audience that Hannah Arendt felt was an important part of constructing a more effective public sphere. This point is demonstrated by the following quote from the EPA report (qtd. in Hobson and Fishman 1998: 4):

> The major intended use is to provide public information to allow community members to request further and more thorough investigations. Results will be communicated through the local media, community meetings, and other methods for distributing information at a neighborhood level. The project expects that community members will use this data at their own discretion, making follow-up requests with community groups, government agencies and facilities on the basis of this data and drawing on other sources for information about the health effects of pollution levels. The project does not intend to use the data for any regulatory, medical, or legal uses for the duration of the pilot project.

Although not ruling out the possibility of more direct regulatory use of citizen data in the future, this case illustrates an agency engaging citizen science in a manner that permits educating the public, broadening the audience, and enabling citizens to employ the data to achieve informal social control.

Integrating Citizen Science into the Legacy Act

The Legacy Act offers several significant opportunities to encourage citizen volunteers to participate in the collection and diffusion of environmental data and thereby creates opportunities for the publicity necessary to cultivate a more active and informed citizenry. This section summarizes the opportunities the Act presents and addresses some of the issues raised by adapting the bucket-brigade lessons to the Legacy Act context.

As noted previously, primary roles for incorporating citizen science under the Legacy Act are for collection of baseline data (section 5) and ongoing monitoring of resources (section 10). The citizen suit provision provides a further tool that provides leverage and opportunity for citizens to engage resources and create partnerships with NGOs and academics. Similarly, as described previously, a provision authorizing and directing stewardship agencies to support education of citizen scientists and development of data-collection protocols and to perform quality assurance on data would be a valuable addition that would enhance data quality.

Although the bucket-brigade experience offers a useful model, it is important to note the similarities and differences presented by employing citizen science in the Legacy Act context. Some activities such as water-quality monitoring of water resources on public lands might be similar in many ways to the air-quality monitoring performed by bucket brigades. Indeed, the cost of water-quality-monitoring technology and testing may make water-quality sampling even more affordable. Another example where citizen participation might present no more obstacles than air-quality monitoring by bucket brigades would be citizen observation and documentation through photography and videography of degrading activities on public lands. Citizens already play that role under existing law. For example, citizens played important roles observing and documenting the degradation of public lands in Utah by off-road vehicle use. Similar observation and documentation could have an even more powerful effect under the Legacy Act.

However, data collection on other types of resources, such as the state of biodiversity, is likely to impose different demands that citizen-scientists might not always be able to meet. For example, monitoring of biodiversity can require more systematic and extensive data collection than does onetime collection of air- or water-quality samples. It is still possible that citizens could be trained to play a role under scientist supervision, but this is not certain. Another important distinction from the bucket-brigade experience is the fact that, apart from wetlands and endangered species, legacy resources are found largely on public lands. In the case of bucket brigades, citizen engagement arose from pollution surrounding and infiltrating peoples' homes and neighborhoods – where they lived. This raises the question of whether citizens would be motivated to participate in monitoring of natural resources on public lands. However, today local groups exist dedicated to protecting regional natural areas, as do citizen support organizations formed to support federal public lands such as national parks. This suggests that concerned and motivated citizens are willing to devote their time and energy to protecting public lands. These same citizens could be engaged with greater effect under the Legacy Act.

Conclusion

The Legacy Act is an opportunity to break away from the passive participation models of prior U.S. environmental law. A potential way to increase citizen participation is to involve citizen science groups in the collection of better environmental baseline data and monitoring resources over time. In the collection of data and the public distribution of stories about the status of public natural resources, citizen environmental monitors could increase the amount of information about the ecological condition of local resources, build alliances, and networks of concerned citizens in the community. The broader array of citizens and citizens' groups could help build a broader base of support for conservation of public natural resources and bring about more actors deploying informal means of social control. In this way, they help broaden the witnesses, narrators, and audiences that Arendt felt were instrumental in an effective public sphere or democracy. Citizens' monitoring groups embody the activity, speech, and disclosure of accounts to an audience (regulators, scientists, citizens, neighbors) in ways that Arendt would recognize as important forms of publicity and thus a more effective public sphere.

Certainly, Arendt would argue for more face-to-face interaction among witnesses, narrators, and their audience; however, Arendt also realized that publicity can happen in degrees. In a world of complexity with myriad local production processes and environmental impacts, the new generation of environmental law can help produce local spheres of publicity about site-specific local environmental problems.

REFERENCES

Ackerman, Frank, and Lisa Heinzerling. 2004. *Priceless: On Knowing the Price of Everything and the Value of Nothing. New York: New Press.*

Arendt, Hannah. 1958. *The Human Condition*. Chicago: University of Chicago Press.

Arendt, Hannah. 1982. *Lectures on Kant's Political Philosophy*. Edited by Ronald Beiner. Chicago: University of Chicago Press.

Brown, Phil. 2007. *Toxic Exposures: Contested Illnesses and the Environmental Health Movement*. New York: Columbia University Press.

Center for the Study of the American Electorate. 2006. *2006 Primary Turnout a Record Low – 15 Percent of Eligibles Vote*. Washington, D.C.: Center for the Study of the American Electorate, American University.

Center for the Study of the American Electorate. 2008. *2008 Primary Turnout Falls Just Short of Record Nationally, Breaks Records in Most States*. Washington, D.C.: Center for the Study of the American Electorate, American University.

Eden, Sally. 1996. "Public Participation in Environmental Policy: Considering Scientific, Counter-scientific and Non-scientific Contributions." *Public Understanding of Science* 5(3): 183–204.

Environmental Protection Agency. 1979. Quality Assurance Policy Statement. January 24, 1979. http://www.epa.gov/ttn/nsr/psd1/pdf/p7_2.pdf.

Environmental Protection Agency. 1998. *Introduction to the National Directory of Volunteer Environmental Monitoring Programs*, 5th ed. Washington, D.C.: Environmental Protection Agency.

Environmental Protection Agency. 1999. "National-Scale Air Toxics Assessment." http://www.epa.gov/ttn/atw/nata/1999/natafinalfact.html (last updated November 6, 2007).

Environmental Protection Agency. 2008. "Technology Transfer Network Air Toxics Web Site," http://www.epa.gov/ttn/atw/nata1999/natafinalfact.html.

Hobson, Jeff, and Schuyler Fishman. 1998. "Community for a Better Environment Quality Assurance Program Plan for the 'Bucket Brigade' Community Air Sampling Pilot Project," http://www.gcmonitor.org/downloads/QAQCPLAN.PDF.

Kleinman, Daniel Lee, ed. 2000. *Science, Technology, and Democracy*. Albany: State University of New York Press.

Laasby, Gitte. 2007. "Better Air Testing Demanded," *Post-Tribune of Northwest Indiana*, August 3, A13.

Louisiana Bucket Brigade. 2009. "Norco: Profile," http://labucketbrigade.live.radicaldesigns.org/article.php?list=type&type=23.

May, James R. 2003. "Now More Than Ever: Trends in Environmental Law Suits at 30." *Widener Law Symposium* 10: 1–52.

McCombs, M. E., and D. L. Shaw. 1972. "The Agenda-Setting Function of Mass Media." *Public Opinion Quarterly* 36: 176–87.

McGarity, Thomas O. 2008. "Hazardous Air Pollutants, Migrating Hot Spots and the Prospect of Data-Driven Regulation of Complex Industrial Complexes." *Texas Law Review* 86: 1445–92.

Nijhuis, Michelle. 2003. "Passing the Bucket," *Grist*, July 23.

Overdevest, Christine, and Brian Mayer. 2008. "Harnessing the Power of Information through Community Monitoring: Insights for Social Science." *Texas law Review* 86: 1493–1526.

Overdevest, Christine, Cailin Huyck Orr, and Kristine Stepenuck. 2004. "Volunteer Stream Monitoring and Local Participation in Natural Resource Issues." *Human Ecology Review* 11: 177–82.

Putnam, Robert D. 2000. *Bowling Alone: The Collapse and Revival of American Community*. New York: Simon & Schuster.

Sapong, Emma D. 2005. "Two Plants Accused of Air Pollution." *Buffalo News Northtowns Bureau*, February 17.

Skocpol, Theda. 2003. *Diminished Democracy: From Membership to Management in American Civic Life.* Norman: University of Oklahoma Press.

St. Croix, G.E.M. de. 1975. Political Pay outside Athens. *Classical Quarterly* 25(1): 48–52.

Strickland, Eliza. 2004. "To Clean the Air, Communities Grab a Bucket," *Christian Science Monitor*, April 1.

Tocqueville, Alexis de. 2000. *Democracy in America*. Translated by H. Mansfield and D. Winthrop. Chicago: University of Chicago Press. (Originally published in 1835.)

6 Creating National Environmental Legacy Act Information

The Double Standard

Walter A. Rosenbaum

> In the scientific and technical world, information and data are commonly judged on their accuracy and precision, repeatability, and, more generally, the degree to which they truthfully reflect conditions in the real world. . . . In political and policymaking circles, a primary criteria for good information is relevance – does it deal directly with the question at hand, does it illuminate the issue in way that points to a solution?
>
> Robin O'Malley, Kent Cavender-Barnes, and William Clark, *Environment Magazine*, May 1, 2003

THE NATIONAL ENVIRONMENTAL LEGACY ACT (LEGACY ACT OR Act) would create a profound transformation in national resource policy if it did no more than declare a presumptive, clearly enforceable standard for the nation's future resource management. However, implementingthe Legacy Act's substantive goals also requires the utilization of a rich and, in many respects new, information base that must be sufficiently refined to be accessible to resource managers at many government levels. The Legacy Act is grounded in the development and refinement of this distinctive data foundation. State and local governments, as well as federal resource managers, will be substantially challenged by the innovations inherent to this essential data resource. In the end, the data required by and developed pursuant to Legacy Act should meet two standards: it should be scientifically and technically credible, and it should be policy relevant.

This chapter examines how Legacy Act can meet these goals. It begins by examining three significant issues in the design and implementation of

Legacy Act's data provisions: the process used in defining the substantive legacy, Legacy Act's approach to uncertainty, and the design and implementation of Legacy Act's metrics and data requirements for measuring resource baselines and ongoing monitoring of resources. It considers both the potential pitfalls and opportunities posed by the design of these information provisions. Taking a broad perspective that acknowledges the important interrelationships among federal, state, and local government efforts to conserve natural resources, it then explores the lessons that may be learned from state and local government experience and the potential downstream impacts that Legacy Act information demands may have for these governments.

I. A Trinity of Data Fundamentals

A. The Importance of Process in Developing a Modest Standard of Sustainability and Appropriate Metrics

Process design is as essential as substantive goals in the development and implementation of the Legacy Act's substantive standard and in its implementing data requirements. Process is a recurrent theme in the discussion of data utilization because controversies over the adequacy of resource stewardship are often grounded in disagreement over the validity of stewardship measures After a comprehensive review of federal, state, and local experience with developing environmental indicators – a fundamental component for legacy management planning – the congressional Government Accountability Office (GAO) cautioned that "the process of developing an indicator set can be an intensely political process that challenges both the credibility and relevance of a set."[1] To avoid the stigma of political bias, the GAO emphasized that developers of data sets "largely relied on collaborative processes to define the purpose and intended use of an indicator set, determine the conceptual model and criteria for selected indicators, and selecting the indicators themselves."[2] Thus, a carefully crafted process for defining management goals, selecting resource indicators, and establishing monitoring procedures throughout a legacy period may at least diminish, if not preclude, subsequent implementation conflicts. This process should be open, stakeholder inclusive, and scientifically informed as well as administratively practicable.

The transcendent goal of the Legacy Act is "to determine a threshold of resources that we commit to leave to future generations, and [to ensure] that necessary savings will occur.... the Legacy Act requires stewards of public resources to set aside a defined level of resources for future generations, and to spend or use only that which remains or can renew itself."[3] This is intended to be the clear, enforceable standard that distinguishes Legacy Act from other federal resource management goals.

Whatever the ultimate standard, the goal of enacting the Legacy Act is inherently linked to the notion of sustainability. In defining the chosen legacy, Congress will implicitly define a goal that bears some relation to this concept. Resource managers will then have to come to terms with the meaning of the concept of sustainability, translating this conception into a metric relevant to a specific resource and its measured use over time. In the end, a resource agency's compliance with the Act will depend less on the volume of documentation validating a particular resource management goal than on a legacy plan that permits agencies to determine whether proposed activities will entail depletion in excess of the statutory standard. Absent sufficient proof, activities will be barred.

In 1984, *Our Common Future* (the "Brundtland Report") proposed a now-pervasive conception of sustainability as "meeting the needs of the present without compromising the ability of future generations to meet their own needs," and thereby initiated a sustained debate concerning how to translate these ambiguities into practicable, policy-relevant terms.[4] This discourse will arise inevitably in the process of the Legacy Act's implementation. The Act itself does not explicitly mention sustainability in characterizing its statutory standard, but a modest idea of sustainability is implicit to the standard's conception. While acknowledging that the ultimate choice of a standard must be made through an accountable political process, Professor Flournoy offers an illustration of what a Legacy Act standard might look like. She proposes "a standard that prohibits any degradation or depletion of covered renewable resources over the legacy period."[5] Any plan to protect a resource by preserving a baseline over a legacy period implies some determination of which among many possible variables will go into the calculus.[6] For example, should resource managers consider only ecological services, or should they also factor in social and economic services in setting desirable maintenance levels for a resource? How ought these different factors be weighted? Who should determine cloture on the list of variables to be considered in estimating the sustainability of a particular resource?

In this endeavor, the experience of state and local governments may offer valuable experience and guidance for the process by which Congress develops the substantive standard. As a practical matter, the method used to establish management goals and appropriate indicators for a specific resource – which entails sorting out the mix of variables to include in the final sustainability calculus – is likely to be de facto negotiated among stakeholders; however, it may eventually be declared de jure. This implies the importance of early establishing a transparent, publicly accessible process not only when creating an information base but also when determining specific resource goals over a legacy period.

State and local governments crafting sustainable resource management plans are finding the existence of such a process almost a prerequisite for success. The experience of Oregon, an early leader in state sustainability planning, is instructive. The Oregon Sustainability Board, created by the Oregon legislature in 2001 to promote sustainable state and local development, had an early cautionary experience: "Currently no significant institutional capacity exists or is available to convene business, government and nonprofits on an ongoing basis for the purpose of identifying and assisting on-the-ground action," it noted. "A system to allow government, business, non-profits and citizens to work together more effectively and efficiently would enhance the ability of communities to create sustainable solutions."[7]

This process can involve some risky contingencies. The selection of resource indicators is likely to evoke controversy among different stakeholders promoting competing metrics, indicators, and data requirements. Local public officials and resource managers may be wary of a process that might produce data implying deficient management or oversight. However, the development of the Sustainable Seattle and the Oregon Benchmarks projects – frequently cited examples of innovative state and local sustainable resource planning – suggests that an open and public discourse over indicator formulation can facilitate planning in several ways: it can identify inherent conflicts among stakeholders, attract broad institutional participation within the affected public, and suggest important linkages among indicators that might otherwise be ignored.[8]

B. Coping with Uncertainty: The Precautionary Principle

The illustrative language provided for section 3 of the Legacy Act is intended to prohibit any action that *may* cause impermissible degradation or depletion of the legacy resource. The statute also mandates that any doubt created by inadequate or incomplete information should be resolved by protecting the legacy, an especially important imperative. Whatever the precise standard that might be selected by a Congress enacting a Legacy Act, it is essential that such a statute reverse the burden of proof to ensure that uncertainty does not undermine legacy goals. The inevitably constrained information base and contested procedures for establishing sustainable levels of resource use, notwithstanding our best efforts to design workable metrics and to generate necessary data, ensure that a substantial measure of uncertainly will attend any effort to assess whether a particular action may result in unacceptable impacts on a specific resource. Therefore, the Act places the burden of proof that such activity will not degrade a resource on those who propose degrading or depleting action.

This constitutes an application of the precautionary principle, still uncommon in risk management among U.S. governmental agencies, as a decision rule, The essence of the precautionary principle is embodied in the so-called Wingspread Statement, a frequently cited definition created in 1998 by a group of scientists, environmental activists, government officials, lawyers, and labor representatives at Wisconsin's Wingspread conference center: "When an activity raises threats of harm to human health or the environment, precautionary measures should be taken even if some cause and effect relationships are not fully established scientifically. In this context the proponent of an activity, rather than the public, should bear the burden of proof."[9]

This is a very different logic from the risk management strategies traditional to U.S. regulatory culture.[10] The precautionary principle is widely used by the European Union and its member states, but very few U.S. state and local governments have any experience with it. Instead, U.S. environmental regulations commonly place on regulatory agencies the burden of characterizing environmental harm from an activity, of estimating the likelihood of such harm, and of demonstrating that the anticipated harm exceeds permissible regulatory limits.

The Legacy Act's implementation of a version of the precautionary principle does not necessarily require the creation of new resource data, but it does require resource planners to evaluate the available information in an unfamiliar way. For example, a precautionary approach to resource planning under the Act presumes some procedural guidelines for resource managers when evaluating the potential adverse impact of an action on a legacy resource:

- *Alternatives assessment*: An obligation to examine a full range of alternatives and select the alternative with the least potential impact on human health and the ecological systems, including the alternative of doing nothing.
- *Anticipatory action*: A duty to take anticipatory action to prevent harm.
- *Right to know*: A responsibility to ensure that relevant public has complete and accurate information on potential human health and environmental impacts associated with the selection of products, services, operations, or plans.
- *Full cost accounting*: A duty to consider all the costs, short- and long-term, implied by a determination of resource management over a legacy period.
- *Due diligence*: A responsibility to ensure that the decision process is transparent, participatory, democratic, and informed by the best-available independent science.

C. Creating Environmental Indicators, Baselines, and Monitoring

Baseline data and metrics are the foundation for evaluating legacy stewardship. The Legacy Act's implementation is data intensive in its requirement for a technically credible, updated base of metrics or indicators appropriate to determinations of resource sustainability over legacy periods. These indicators, in turn, become the empirical currency for creating baseline data and legacy plans – creating a historical record of critical resource attributes and projecting the measurements through the legacy period. If, for example, wetlands preservation is a legacy objective, the availability of soil- and water-quality indicators (and determining which indicators to monitor) becomes essential.

Indicators should have several qualities:

- Linkage to clearly defined goals and objectives derived from conceptual models based on key resource structures and functions;
- A structured, logical process for selection and development that is scientifically rigorous and transparent; and
- Stakeholder agreement on the indicator selection criteria.

The federal government, state and local governments, numerous foundations, and nongovernmental organizations (NGOs) are creating or acquiring environmental indicators and baselines.[11] Many of these collections, though often historically or spatially limited, are amenable to adaptation for Legacy Act–related legacy planning, or at least for providing a data foundation for such planning. As a practical matter, however, most expenditures on the acquisition of environmental information have been related to regulatory enforcement (e.g., the monitoring of water-quality indicators required by the Clean Water Act) rather than to Legacy Act–relevant sustainability variables. Indicators of legacy stewardship, for instance, might include socioeconomic measures or indicators of cumulative impacts from selected activities affecting a resource over a legacy period.[12]

The Legacy Act also assumes continuous monitoring of resource conditions throughout a legacy period. From an administrative perspective, this implies agreement among legacy managers and stakeholders concerning how monitoring will occur, when it is to occur, and which indicators will constitute the monitoring inventory. Again, process becomes important because the task of resource monitoring may be shared among resource managers and other stakeholders who need to reach agreement on a monitoring strategy – currently, for instance, voluntary civic organizations often assume major monitoring responsibilities in resource management and might do so here (see Chapter 5 in this volume).

II. The Policy Relevance Test

In the end, regardless of how otherwise meticulous the creation of a Legacy Act database, the Act should be designed to speak to policy makers in a coherent and relevant language. "Ideally," note the designers of the Heinz

Center's widely respected *The State of the Nation's Ecosystems, 2008*:

> [P]olicymakers confronted with an emerging issue would also be able to call on a set of facts that describe the most important aspects of the issue at hand, are perceived by the relevant technical experts to be doing so accurately, and are not perceived as having been chosen so as to benefit one position or program and damage another.[13]

"Relevance" also implies a language of resource characterization written in terms accessible to decision makers and stakeholders, most of whom will not be technical specialists. This does not imply dumbing down by eliminating important data nuances and qualifications. It does imply a deliberate effort at conciseness and alertness to the interpretative skills of the prospective audience, which is likely to include not only governmental officials and technical experts but also numerous representatives of private-sector advocacy groups and the public. Moreover, considerable discontinuities still exist, especially at the local governmental level, between the increasing technical sophistication of available environmental data and the interpretative skills of relevant decision makers, especially local elective officials. Data formatted according to a geographic information system (GIS), for instance, is widely available and often downloadable from monitoring agency Web sites in many states but infrequently used by elective resource managers. The concept of metrics proposed by Steinzor and Shapiro[14] and adopted in the Legacy Act design seeks to bridge this gap, but much work will need to be done to identify the appropriate metrics.

From their experience in creating national indicators of environmental quality, participants in the Heinz Center suggest several criteria applicable to formatting data for Legacy Act decision makers:

- *Policy relevance*: it must answer questions that are or are likely to be the subject of policy action of debate.
- *Technical credibility*: the relevant scientific or technical experts must perceive the information as consistent with currently accepted scientific norms.
- *Political legitimacy*: there must be no hint of agenda-driven selection, or skewing, or spinning the data, and the presentation must be rigorously nonpartisan.[15]

In effect, the relationship of information manager and decision maker can be compared to a data marketplace in which information effectively consumed by the appropriate decision makers must be packaged attractively for potential customers. All this involves anticipatory planning and vetting of databases before their use in Legacy Act decision making. The director of the Northwest Environmental Watch (NEW), an early Washington State indicators project, recommended sensitivity to the policy context in which the resource decision maker must act as a means of targeting indicator selection:

> NEW recognized that their project must "tell [targeted readers] something they care about" and "be somehow connected to action." NEW recognized that policy makers, upon reading an indicator, immediately want to know how the indicator can be applied to a policy context in order to maximize the performance of that policy-maker's office.[16]

Targeting Legacy Act data in this way implies early deliberation with policy makers in the process of data design, creation, and monitoring. Moreover, it implies negotiations not only between policy consumers and information producers but also among scientific and technical specialists to reduce the number of potential indicators to a coherent and understandable core – a strategy that will almost inevitably require some hard-won compromise among the experts. This translates into clear signals on the importance of some vehicle akin to the committee of experts recommended for Legacy Act[17] but suggests the need for an additional and broader process that includes representatives of agencies, independent experts, and interested public and private stakeholders. Some type of two-tiered process with a deliberative scientific body and a policy-focused body would have the benefit of ensuring the technical credibility as well as political legitimacy and policy relevance of the indicators chosen.

III. The Open Public Window

The resource management goals of the Legacy Act may be inherently important to the public, but this is not necessarily obvious, particularly in light of the challenge of focusing public attention on the problems of the future and intergenerational issues. Skillfully crafted Legacy Act information management with an eye to public and media interest can create a potent constituency for the Act's goals and evoke the participation of important institutional interests in the Act's implementation. A place needs to be created early for media and public involvement in the design and collection of Legacy Act data as well as attention to the Act's implementation.

Ongoing state and local environmental indicator projects commonly pay attention to media investment in the process and public awareness of its significance. "Indicator projects have the potential to report data in ways that are equally compelling to scientists who understand what was done to compile the data and journalists and members of the public who understand a compelling story," concludes a careful evaluation of Seattle's continuing indicators project.[18] Moreover, "the creation and use of indicator projects by the public and civil society groups have the potential to challenge commonly-held beliefs with powerful, data-driven arguments."[19] Another implicit advantage in opening the data management process to public and media view is that it encourages (or chastises) the relevant scientific experts, administrators, and elective policy makers to cultivate an information language that communicates effectively to laypeople.

IV. The Diffusion of Legacy Act Innovation

If the Legacy Act proves administratively and politically attractive among federal resource managers, its institutional evolution may continue among public resource agencies. Although the Legacy Act is intended to be a strategic guide for national resource managers, its stewardship goals and management assumptions could permeate throughout the governmental system. States could enact their own state specific legacy acts, replicating the history of the National Environmental Policy Act (NEPA) of 1969, which began as a federal statute and subsequently evolved into little NEPAs being enacted by fifteen states and the District of Columbia within a decade of NEPA's creation. In addition, under the cooperative federalism that characterizes some aspects of our natural resource management laws, state agencies may need to coordinate their own data collection with the Legacy Act's data standards to participate fully. State and local governmental access to resource-related federal grants or other appropriations for resource management may be contingent on the capacity of those governments to assess resource impacts in terms of the Legacy Act's substantive goals and data requirements. Thus, even without state legacy acts, state and local governments may feel effects from the enactment of legacy act.

In designing the Legacy Act provisions related to data demands, metrics, and monitoring, Congress should draw not only on federal agency best practices and experience to date but also on the information available through state and local government associations that create and disseminate information about best practices in numerous policy domains, including resource management. These best-practices materials, promoting innovative solutions to such problems as data acquisition, planning guidelines, and public involvement, often become a primary pathway for the diffusion of innovation among local governments. An increasing number of national and international organizations have created accessible databases and guidelines for use by state and local governments concerned with sustained resource use such as the Legacy Act promotes.[20] The Global Reporting Initiative (GRI),[21] for example, offers local governments accessible data sets of local environmental indicators, training in the development of local indicators, and guidance in the display and interpretation of indicators. The Community Indicators Consortium, a collaboration among local governments, sponsors annual conferences on indicator development and use.[22]

The Legacy Act often rubs against the grain of U.S. experience with resource management and risk assessment. Most state and local resource agencies have little, if any, experience in establishing an explicit, scientifically defensible resource management goal premised on sustainability or in crafting the implementation guidelines. For instance, the Act requires consideration of anticipated cumulative impacts of legacy-related activities and a careful assessment of the capacity of renewable resources to replace themselves – neither indicator is common to even such widely used documents as NEPA's *Environmental Assessments*. Thus, the diffusion of a Legacy Act standard for resource management throughout the federal system will also require considerable institutional innovation. The substantive goals of the Act and the requisite information to implement them entail a substantial challenge in data management for all levels of government – national, state, and local – especially, a significant amount of learning for both professional staff and elective officials. Moreover, environmental data sharing among state and local governments is not robust, and opportunities should be created to promote Legacy Act–related discourse among resource agencies.

The challenges will be especially acute for the many local governments lacking resources or experience in new, rapidly involving information formats such as GIS and its variations. The California Environmental Protection Agency's *Environmental Indicators for California* (EPIC), among the most ambitious and innovative emerging state databases, has adopted GIS formatting as a fundamental principle.[23] In some instances, the Legacy Act's implementation will assume the availability of a data cache that does not yet exist or may require the evaluation of information through expert interpretation, perhaps not readily accessible. The concern of the Act for transparency in data management mandates that baseline information and updated monitoring data be made publicly accessible, perhaps on the stewardship agency's Web site.[24] Local governments vary enormously in their ability to respond effectively to such imperatives.

Conclusion: Meeting the Double Standard

The Legacy Act demands supporting data sufficient in quantity to provide information technically and scientifically relevant to the Act's many resource management goals and crafted so as to be accessible and relevant to policy makers. The effectiveness of the Act will depend on this information and the process by which that information is conceived, obtained, and validated. Thus, meeting the double standard of scientifically appropriate and policy-relevant information will require the Legacy Act's proponents to invest considerable time and care in the design of the information base on which it will be grounded.

Experience with the development of indicator data and other information relevant to sustainable resource management at the state and local level suggests that good process design is likely to significantly facilitate the Act's implementation perhaps as much as the availability of good data itself. Good design also takes time, consumes significant resources, and requires a significant commitment of leadership talent from the outset. Yet the potential benefits of such an investment could include not only successful preservation of our desired federal public resource legacy but also enhanced resource management standards and data at the state and local levels.

NOTES

1. GAO, *Better Coordination Is Needed to Develop Environmental Indicator Sets Thlat Inform Decisions,* November 10, 2004, GAO-05-52, at 5. Washington, D.C.
2. *Id.* at 5.
3. Chapter 1 in this volume.
4. World Commission on Environment and Development, *Our Common Future* 43 (New York: Oxford University Press, 1987).
5. *See An Enforceable Substantive Mandate (Sections 1 and 3),* Chapter 1 in this volume. Recognizing the more complex tradeoffs involved with nonrenewable resources, Flournoy proposes a standard that prohibits significant degradation or depletion.
6. For example, the number of potential variables that might be considered in choosing a resource indicator include cost-effectiveness of the acquisition, its relative importance to decision makers, its reliability, its temporal or spatial scale, and its robustness among many others. Minnesota's Department of Natural Resources provides some useful guidelines in *Using Indicators of Natural Resource Development: DNR Indicator Development* (http://www.dnr.state.mn.us/eii/use.html).
7. Oregon, Sustainability Work Group, The Sustainability Work Group Initial Report, November 2000 at/www.sustainableoregon.net/govt/group.cfm (October 18, 2009).
8. Steve Connelly, "Mapping Sustainable Development as a Contested Concept," *Local Government* 12, no. 3 (2007): 259–78; Michael S. Carolan, "Sustainable Agriculture, Science and the Co-production of 'Expert' Knowledge: The Value of Interactional Expertise," *Local Government* 11, no. 4 (2006): 421–31; Marc Schlossberg and Adam Zimmerman, "Developing Statewide Indicies of Environmental, Economic and Social Sustainability: A Look at Oregon and Oregon Benchmarks," *Local Government* 8, no. 6 (2003); Meg Holden, "Community Interests and Indicator System Success," *Social Indicators Research* 90, no. 1 (August 2008): 641–60; Meg Holden, "Revisiting the Local Impact of Community Indicators Projects: Sustainable Seattle as Prophet in Its Own Land," *Applied Research in Quality of Life*, Vol. 1, No. 1 (2006): 253–77; Hubert L. Rees, Jeffrey L. Hyland, Ketil Hylland, Colleen S.L. Mercer Clarke, John C. Roff, and Suzanne Ware, "Environmental Indicators: Utility in Meeting Regulatory Needs *ICES Journal of Marine Science* 65 (2008): 1381–86.
9. "The Precautionary Principle," Science and Environmental Health Network, January 2000, (http://www.sehn.org/ppfaqs.html).
10. The City of San Francisco is the only municipality or state enacting an environmental ordinance involving the principle: "Where threats of serious or irreversible damage to people or natural systems exist, lack of full scientific certainty relating to cause and effect shall not be viewed as sufficient reason for the City to postpone measures to prevent the degradation of the environment or protect the health of its citizens.... Where there are reasonable grounds for concern, the Precautionary

Principle is meant to help reduce harm." City of San Francisco, *Precautionary Principle Policy Statement*, chap. 1, sec. 101. "The San Francisco Precautionary Principle," Ordinance 171–03, approved July 3, 2003 (http://www.municode.com/content/4201/14134/HTML/ch001.html).

11. Many national indicators can be accessed through the U.S. Environmental Protection Agency's Environmental Indicators Gateway (at http://www.epa.gov/igateway/index.html). The federal government's Council on Environmental Quality, Office of Management and Budget, and Office of Science and Technology Policy recently launched an initiative, National Environmental Status and Trends, requiring federal agencies to develop national environmental indicators and together plan a major indicators pilot project. Many of the most currently developing databases are associated with sustainability goals: U.S. Environmental Protection Agency (EPA), Sustainability (http://www.epa.gov/sustainability), U.S. EPA, National Center for Environmental Research, Estuarine and Great Lakes Program (EAGLES) (http://www.aseinc.org), ICLEI – Local Governments for Sustainability (http://www.iclei.org), Sustainlane Government (http://www.SustainLane.org), and SCN: Sustainable Communities Network (http://www.sustainable.org). Although no longer updated, a still-useful directory of indicator sources is Florida State University, Program for Environmental Policy and Planning, Environmental Indicator Technical Assistance Series (http://www.pepps.fsu.edu/segip/catalog). *See also* Tomas Barros Ramos, "Environmental Performance Indicators for the Public Sector," in *Encyclopedia of Earth*, ed. Cutler J. Cleveland (Washington, D.C.: Environmental Information Coalition, National Council for Science and the Environment, 2008), http://www.eoearth.org/article/Environmental_performance_indicators_for_the_public_sector.

    Many U.S. organizations are developing comprehensive key indicator systems – organized systematic efforts to produce selected economic, social, and environmental indicators – to assess position and progress toward specific goals. *See* GAO, *Informing Our Nation: Improving How to Understand and Assess the USA's Position and Progress*, November 10, 2004, GAO-05–01, Washington, D.C.

12. The difference between measuring environmental regulatory actions and outcomes in resource management and the need to clarify outcomes more clearly is carefully examined in National Academy of Public Administration, *Taking Environmental Protection to the Next Level: An Assessment of the U.S. Services Delivery System* (Washington, D.C.: National Academy of Public Administration, 2007). *See also* U.S. EPA, Region 3, *Mid-Atlantic Water: Tools to Measure Our Progress* (http://www.epa.gov/reg3wapd/indicators/index.htm#indicators).

13. The H. John Heinz III Center for Science, Economics, and the Environment, *Highlights: The State of the Nation's Ecosystems, 2008* (Washington, D.C.: The H. John Heinz III Center for Science, Economics, and the Environment, 2008), 5.

14. Sidney A. Shapiro and Rena Steinzor, *Capture, Accountability, and Regulatory Metrics*, 86 Tex. L. Rev. 1741 (2008).

15. Robin O'Malley, Kent Cavender-Barnes, and William Clark, "Better Data: Not As Simple As It Seems," *Environment*, 45, No. 4 (May, 2003), 10.
16. Meg Holden, "Revisiting the Local Impact of Community Indicators Projects: Sustainable Seattle as Prophet in Its Own Land," *Applied Research in Quality of Life*, 1, No 2 (April, 2006), 264.
17. *See Baseline Information and Metrics (Section 5)*, Chapter 1 in this volume.
18. Holden, *supra* note 16, at 267.
19. *Id.*
20. *See supra* note 11.
21. A major component of the GRI is the Sustainability Reporting Framework that "provides guidance for organizations to use as the basis for disclosure about their sustainability performance, and also provides stakeholders a universally-applicable, comparable framework in which to understand disclosed information. The Reporting Framework facilitates transparency and accountability by organizations – companies, public agencies, non-profits – of all sizes and sectors, across the world" (http://www.globalreporting.org/AboutGRI).
22. The Community Indicators Consortium is committed to "an active, open learning network and global community of practice among persons interested or engaged in the field of indicators development and application. It is organized to: advance the art and science of indicators; facilitate the exchange of knowledge about the effective use of indicators; encourage development of effective indicators; and foster informed civic and media discourse about local, regional, national, and global priorities." (http://www.communityindicators.net/about.html).
23. GIS formatting also facilitates integration with other databases. "To be most useful, environmental indicator systems must take advantage of new scientific knowledge, better analytical capabilities, regulatory changes, new technologies, and adapt to shifting priorities. For example, geographic information systems (GIS) represent a technological tool that will be used to enhance EPIC's ability to evaluate, manage and present indicator information. EPIC will also coordinate its activities with efforts under the Office of Environmental Health Hazard Assessment's Emerging Environmental Challenges Program to identify and characterize issues that may confront the state in the future updates." California Environmental Protection Agency, *Environmental Indicators for California*, "Executive Summary."
24. *See, e.g.*, King County (Washington), Department of Natural Resources, *KingStat 2007* (http://www.metrokc.gov/dnrp/measures), and Chesapeake Bay Program, *Data Hub* (http://www.chesapeakebay.net/dataandtools.aspx?menuitem=14872).

7 The Constitution and Our Debt to the Future

Rena Steinzor

I. A New Breed of Public Law

CONGRESS GAVE BIRTH TO THE NATION'S MAJOR FEDERAL environmental statutes during a period of intense and extraordinarily fertile social upheaval, as America's young people struggled to reclaim their government and parents fought to recover their children's respect. The first generation of statutes, passed as the Vietnam War was winding down, launched a forty-year revolution in the way Americans treat their environment, propelling unprecedented advancements in pollution control and the preservation of natural resources. Yet somewhere along the line – it is difficult to pinpoint a single event or moment – this progress began to unravel. We learned that the environment of our country was irrevocably linked to the global environment and that frightening changes were under way in the atmosphere as an overload of fossil fuel and other emissions disrupted the planet's climate. The developing world was intent on catching up with the United States and Europe economically but lacked the regulatory infrastructure to moderate the impact of industrialization. Our country backed away from global leadership on environmental issues because energy producers convinced politicians that this role cost too much.

These fateful decisions could not have come at a worse time, as we are belatedly beginning to realize. The world confronts accelerating climate change, an environmental crisis that makes efforts to conquer previous challenges look like mere dress rehearsals. So much has changed – globalization of business, invention of the Worldwide Web, discovery of the human genome – and yet so much has remained the same: disillusionment with government, the false dichotomy of jobs versus environment, the overpowering resistance of corporations to regulation. We strain to find ways for our weakened government institutions – from Congress to the president to the career civil service – to steer the nation out of these blind alleys and back on to the high road.

As the chapters in this volume argue, the reforms necessary to meet these grave challenges must go "beyond environmental law" to a conceptual plane where even the most basic and routine assumptions are revisited. To re-create the atmosphere of revolutionary change that gave birth to modern environmental protection, we must push beyond incremental tinkering. New ideas must be incubated, embraced, enacted, and implemented. This chapter addresses the threshold question in American law: how should we read the U.S. Constitution to justify these new breeds of environmental law? In a nutshell, I urge reexamination of the historical interpretations of Congress's constitutional authority to protect the environment. Instead of justifying federal intervention solely as a product of the national government's interest in fostering interstate commerce, Congress should invoke its authority to safeguard and promote the general welfare of the nation.

The National Environmental Legacy Act (Legacy Act or Act) proposed by Professor Alyson Flournoy in the first chapter of this book epitomizes a new generation of law because it would reject actions that would make humankind's consumption of natural resources unsustainable over the long term. Her proposal would prohibit any use of federally owned property unless the full gamut of natural resources that exist on that land could be replenished in time for future generations to use them. Activities on privately owned land that could destroy natural resources on federal property might also be prohibited. Professor Flournoy proposes a multidecade horizon for such evaluations of sustainability, with the consumption of natural resources reevaluated on an enduring basis through an iterative cycle of decision making.

To the extent that the Legacy Act would require federal managers to conserve natural resources owned by the government, the proposal would fall under the Constitution's grant to Congress of explicit authority to control what happens on federal lands in article IV, section 3, which states that it "shall have Power to dispose of and make all needful Rules and Regulations respecting the Territory or other Property belonging to the United States."[1] But to fully guarantee the sustainable use of federally owned natural resources would require a crucial addition step not explicitly addressed by Professor Flournoy: prohibiting activities on nonfederal land if those uses would violatethe Act's mandates. That extension of the Legacy Act would necessitate a broader search for constitutional authority.

This chapter argues that a reexamination of the constitutional grounding is essential as we design and enact the next generation of environmental law. As the political scientist Terry Davies has observed in the context of responding to another next-generation problem, the regulation of nanotechnology,

> [the current system] was designed to deal with the problems of steam engine technology in the context of a pre-computer economy. It was based on assumptions that most problems are local, that programs can be segmented and isolated from each other, that technology changes slowly and that all important problems have been identified. All of these concepts are no longer valid, if they ever were.[2]

Grappling with these kinds of problems will require extracting ourselves from the outmoded framework of the existing constitutional foundation for environmental law – namely article I, section 8, of the Constitution, commonly known as the Commerce Clause, which reads: "The Congress shall have Power... To regulate Commerce with foreign Nations, and among the several States, and with the Indian Tribes."[3] The arguably fatal flaw in this section of the document is that the most common usage of the word "commerce" is the "exchange or buying and selling of commodities."[4] As Congress expanded the federal government's reach into areas of public law long dominated by the states, the Supreme Court kept pace, broadening its interpretation of the Commerce Clause to encompass activity that could potentially affect the economy, whether or not those activities actually involved the exchange of money.[5] This jurisprudence may be broad enough at the moment to encompass the Legacy Act and other next-generation efforts, although in recent years, the Supreme Court has begun to put the brakes on its historically expansive approach to the ambit of federal authority in relationship to the states. But as the transactions governed by the law – for example, mining today that will make land use unsustainable in thirty years, deployment of nanotechnology today that will cause disruption of ecosystems a hundred years hence, or carbon emissions that occurred twenty years ago but will trigger climate changes that threaten our children's children – Commerce Clause justifications appear increasingly fragile and even begin to teeter on the edge of falling of their own weight.

Rooting environmental law in the analysis of its economic effects has also weakened its effectiveness to an extraordinarily corrosive extent. The long-standing assumption that the pros and cons of environmental policy must be rationalized in terms of money reached its apex three decades ago, with the militant application of cost-benefit analysis to regulatory decisions. Supporters of America's strong laws reacted instinctively to these developments by attacking that methodology on its own terms.[6] But it did not occur to us to think about whether our tacit acceptance of the Commerce Clause as the constitutional foundation for environmental and other health and safety laws was the genesis of these developments. Although the Clause does not lead inexorably to reliance on cost-benefit analysis, the parallel between the two is striking. Cost-benefit analysis excludes nonmarket values that are not easily quantified, just as the Commerce Clause focuses on the economic implications of a decision to the exclusion of other, transcending concerns, such as the protection of future generations' health and welfare.

At first and maybe even at second blush, my reasoning will appear to push against the strong tide of immutable constitutional theory when reformers of environmental law have much more pressing and immediate work to do. But I am convinced that this apparently quixotic exercise is well worth the effort. The widespread tendency to employ economics as the primary lens through which we justify health, safety, and environmental regulation has turned the missions of the major environmental laws on their heads, compelling us to consider whether the marketplace will address the problem, as opposed to whether and to what extent the government has responsibility for protecting public health and the environment.

Americans count on their government to prevent a growing number of international disasters – from pandemics to global terrorism to water shortages – and do not conceive of these protections as justified only with respect to "free market," economic concerns. Not only did the framers of the Constitution recognize these expectations; they embraced them in article I, section 8, which authorizes Congress to "provide for the general Welfare" by taxing, spending, and making all "necessary and proper" laws.[7] If commerce at its most fundamental level is comprised of the exchange of commodities and money, then the term refers to a "marketplace" at a specific point in time. Yet ideas like preserving natural resources for future generations or otherwise preventing pollution that could harm our children and their children demand a significantly more attenuated frame of temporal reference than has applied to the more immediate interventions that characterize much of environmental law.

II. The Narrowing Future of the Commerce Clause

Historically, the Supreme Court embraced a broad definition of Congress's Commerce Clause authority, keeping step with national political developments that expanded the federal role in every aspect of domestic policy, beginning with the New Deal and reaching an apex in the Vietnam War and Watergate era.[8] The furthest reaches of the test ultimately developed by the Court had two crucial elements: (1) the activities Congress sought to regulate could involve "non-economic" transactions so long as (2) those activities had a "substantial" effect on interstate commerce.[9] The Court was more willing to curtail Congress when it perceived that the federal government was manhandling the states. For example, it outlawed federal attempts to "commandeer" state government resources in *New York v. United States*, a case involving the siting of a low-level radioactive waste facility.[10] But because all major environmental laws afford the states the opportunity to volunteer to assume delegated authority to implement federal regulatory requirements, and sweeten the deal with grants to support those state programs, *New York v. United States* seemed to address an extreme example of federal overreaching without much significance for the future.

The 1995 decision in *United States v. Lopez*[11] shook complacency about the Court's willingness to read Congress's Commerce Clause authority so broadly. The case involved a search for guns in a high school senior's locker to provide evidence for a criminal case under the federal Gun-Free School Zones Act of 1990. In a tense, 5–4 majority opinion, Chief Justice Rehnquist argued that the Act had "nothing to do with 'commerce' or any sort of economic enterprise" and that it did not contain any self-limiting jurisdictional provision ensuring its limited application to activities that substantially affected commerce.[12] Justice Breyer's dissent gave full-throated voice to the liberal justices' view that the opinion was a startling departure from precedent. Chiding the majority for shifting direction on the Court's long-standing and expansive definitions of commerce, he warned that its effort to distinguish between economic and noneconomic activity would not only create turmoil in the lower courts but also involve judges in second-guessing Congress in ways that exceeded their appropriate constitutional role.[13]

Despite these warnings, the dissent's supporters in the academy and in practice read *Lopez* as confined to federal incursions into the traditionally state-dominated realm of criminal enforcement. They argued that if Congress merely took the trouble to include detailed "findings" in each new statute that specified how the conduct at issue would substantially affect interstate commerce, it would satisfy the standards set out by the narrow federalist majority on the Court.[14] But the narrow conservative majority on the Court took another swipe at the problem five years later in *Morrison v. United States*, striking down a statute that gave battered women the right to bring tort suits in the federal courts.[15] This time, Congress had minded its manners, making extensive findings regarding the adverse impact of gender-motivated violence on interstate commerce. Justices Rehnquist, Scalia, Thomas, Kennedy, and O'Connor were not satisfied, holding that the Constitution imposes real limits on Congress's Commerce Clause authority when noneconomic activities such as assault are involved.[16] This conclusion demonstrated a serious intention to curb expansive federal regulation of purely intrastate activities. As Professor Robert Percival has written:

> *Morrison* thus raises the prospect that Congress cannot constitutionally regulate intrastate activity that the Court deems noneconomic in character. This could mean that Congress lacks the power to prohibit endangered species from being killed by activity that is not characterized as economic in nature, such as recreational dirt-biking.[17]

Sharp exchanges over the constitutionality of congressional efforts to extend federal regulatory protections to waters existing solely intrastate underscored these fears. The 5–4 decision in *Solid Waste Agency of Northern Cook County v. United States Army Corps of Engineers (SWANNC)*[18] involved a proposal by a consortium of municipalities to use land containing artificially created ponds as a disposal site for baled solid waste. The consortium received all applicable state and local permits but was denied a federal permit to fill some of the ponds with debris.

Because the sensitive areas were isolated from hydrological systems that crossed state lines, the consortium argued that Congress had no constitutional authority to regulate them. The federal government responded that the wetlands were habitat for migratory birds that were pursued across state lines by millions of hunters and bird-watchers annually, establishing the required nexus with interstate economic activity under the Commerce Clause.

The Supreme Court majority (Justices Rehnquist, O'Connor, Scalia, Kennedy, and Thomas) ostensibly based its decision to overturn the permit on its interpretation of the statute. But it warned that extending the statute to habitat for the birds would "push to the limit of congressional authority" under the Constitution, even going so far as to suggest that, if the federal government asserted any analogous claims of jurisdiction in the future, it would be required to identify the "precise object or activity that, in the aggregate, substantially affects interstate commerce."[19] Excusing federal policy makers from this potentially heavy burden, the majority continued, "would result in a significant impingement of the States' traditional and primary power over land and water use."[20]

In dissent, Justice Stevens (joined by Justices Souter, Ginsburg, and Breyer) accused the majority of undermining federal power under the Commerce Clause to regulate activities that "substantially affect" interstate commerce: "[T]o constitute a proper exercise of Congress' power over intrastate activities that 'substantially affect' interstate commerce, it is not necessary that each individual instance of the activity substantially affect commerce; it is enough that, taken in the aggregate, the *class of activities in question* has such an effect."[21] As with the dissenters, environmentalists widely perceived the case and its close successor, *Rapanos v. United States*,[22] as disasters that gutted Clean Water Act wetlands protections because they enmeshed the Army Corps and other federal officials in lengthy deliberations of whether wetlands were isolated. The constitutional implications of the decision remain unclear, although Congress is considering legislation to broaden the Army Corps' jurisdiction to extend to isolated wetlands regardless of the presence of migratory birds, potentially setting up another Commerce Clause challenge before the Court.[23]

Should the Court decide to attempt a realignment of Commerce Clause precedent to include only demonstrably economic effects, undermining not just environmental law but also many other bodies of law, the work will be arduous and could take many years to accomplish. Not only would such an effort require the Court to craft clear rules separating economic from noneconomic behavior, it could mean distinguishing between behavior that has intrastate ramifications from behavior that has interstate ramifications.[24] As Justice Souter reminded his colleagues in *Morrison*, the Court had a similar misadventure in *National League of Cities v. Usery*,[25] when a similarly narrow majority attempted to distinguish between traditional state functions immune from federal control and circumstances in which states behaved more like commercial actors and were subject to federal regulation. Confusion over the Supreme Court's rules ultimately spawned three hundred incoherent decisions by the lower federal courts. The chaos so alarmed Justice Blackmun that he switched his vote only nine years later, holding with a similarly narrow majority in *Garcia v. San Antonio Metropolitan Transit Authority* that Congress was well within its Commerce Clause authority when it regulated the labor conditions of state and local workers.[26]

At the moment, the Court is evenly balanced between conservatives (Justices Roberts, Scalia, Thomas, and Alito) and moderate liberals (Justices Stevens, Breyer, Ginsburg, and Sotomayor), with Justice Kennedy most often casting the swing vote. This narrow division makes it difficult to predict whether the conservatives could attract enough votes to continue their crusade against overreaching federal laws. If they decide to persist, however, next-generation statutes like the Legacy Act could prove even more vulnerable to constitutional challenge than more traditional, long-standing statutes. Because the proposals are geared toward protecting the interests of future generations, the intended beneficiaries – natural resources and lives not yet extant – have no current economic or market-based value. No one alive today can engage in transactions – commercial or noncommercial – with beings not yet in existence. Activities that endanger them lack any nexus to commerce as the term is commonly understood, and conversely, Congress lacks any authority to burden present generations with this imaginary debt to the future. Admittedly, parsing the temporal dimensions of federal statutes to ensure that they benefit only people alive today could prove an intellectual exercise that makes debates over intra- versus interstate effects look like child's play. Yet it is easy to imagine Justice Scalia warming to such work.

From this broader perspective, we can discern Commerce Clause ideas as a fault line that will dog all efforts to address emerging environmental problems. Reliance on the federal authority to promote marketplace vigor sets the stage for the supposed trade-off between jobs and the environment and between public health and prosperity. Democrats insist that we can find better ways to protect the environment, therefore sacrificing less economic growth.[27] Republicans claim that environmental regulations waste money, stifle small business, and cripple the nation's competitiveness.[28] The common ground for both arguments is the view that environmental protection and the economy are in a relationship that demands their trade-off against each other. The proposition that we owe it to future generations to preserve natural resources is likely to prove exceptionally controversial. In fact, opponents will argue that, when we enhance monetary wealth by consuming resources, we provide the economic resources needed by future generations to buy their way out of any irreversible environmental trouble. Or, to phrase the argument another way, resources in and of themselves have no value until they are bought and sold. Refraining from buying and selling them has absolutely nothing to do with encouraging the promotion of commerce, which the national government has always done by stabilizing markets, not by enforcing fanciful prohibitions against market transactions.

Professor Flournoy refers to these arguments as the spend-down ethic, explaining that they implicitly reject any moral or ethical commitment to preserve resources for future generations (see Chapter 1 in this volume). Instead, this ethic posits that ownership of resources and control over how rapidly they are consumed are the sole prerogative of those now living on the planet. Under Professor Flournoy's analysis, unless we take decisive action to fundamentally modify our patterns of consumption, the spend-down ethic will win the day. And it is difficult to articulate an economic reason for making ourselves uncomfortable. Rather, the motivation must come from an ethical sense that we must be responsible stewards for our children's future. And, as it turns out, the framers had similar ideas in mind when they gave Congress aspirational, or affirmative, authority.

III. Safeguarding the General Welfare

> We the People of the United States, in Order to form a more perfect Union, establish Justice, insure domestic Tranquility, provide for the common defence, promote the general Welfare, and secure the Blessings

> of Liberty for ourselves and our Posterity, do establish this Constitution for the United States of America.
>
> – Preamble to the U.S. Constitution

> The Congress shall have power to lay and collect taxes, duties, imposts and excises, to pay the debts and provide for the common defense and general welfare of the United States; but all duties, imposts and excises shall be uniform throughout the United States.
>
> – Article 1, section 1, U.S. Constitution

A. The States' Police Power

The argument that the federal government's efforts to deal with long-term threats to public health and the environment should be lifted from the constraints of the Commerce Clause and placed under the umbrella of the national government's efforts to promote the general welfare is best supported by the Supreme Court's line of cases confirming the states' police power to combat comparable threats. At the time of the American Revolution, this concept was captured in the Latin phrase *salus populi supreme lex est*, or "the safety of the people is the supreme law." States have a long and noble history of regulating practices that could threaten public health, beginning as early as the 1700s, when smallpox inoculations were common in the New England colonies, extending through the sanitarian movement in America's major cities during the late nineteenth and early twentieth centuries, covering the campaign to eradicate polio during the 1950s and up to the present day when the threat of AIDS has challenged their capacity to the breaking point.[29]

The *Slaughter-House Cases* decided by the Supreme Court in 1872 typify this jurisprudence.[30] At issue was a Louisiana law granting a monopoly to a single slaughterhouse on the grounds that it would be easier to control the practices that led such places to spread disease through the careless disposal of animal carcasses and other wastes. The Court treated the case as a showdown between state police-power prerogatives and the recently enacted Fourteenth Amendment grant of "privileges and immunities" to citizens of the United States[31] – in this instance, the chosen slaughterhouse's competitors and their privilege to continue in business. Owners of competing slaughterhouses challenged the law, alleging that their privileges and immunities were violated by the state's grant of a monopoly to a competitor. In response, the Supreme Court cited the "power here exercised by the [Louisiana] legislature," which "in its essential nature" has always belonged to the states."[32] The Court continued:

> This is called the police power: and it is declared by Chief Justice Shaw that it is much easier to perceive and realize the existence and sources of it than to mark its boundaries, or prescribe limits to its exercise.... Upon it depends the security of the social order, the life and health of the citizen, the comfort of an existence in a thickly populated community, the enjoyment of private and social life, and the beneficial use of property.[33]

The *Slaughter-House Cases* established the premise that the states are presumed to have broad police powers and that the question for the federal courts is whether they have gone too far in exercising those. But the strength of this holding weakened over time both because the state law at issue came to be viewed as a cynical exercise in post-Reconstruction patronage and because the threat of epidmics and other urgent public health emergencies waned.

Once the urgency and barely contained terror of cholera and similar diseases was brought under control by better sanitation and medical breakthroughs such as vaccination, state public health officers turned to what Professor Wendy Parmet calls "endemic" threats to public health, such as working conditions and occupational exposure to harmful substances.[34] The federal courts balked, largely because the new targets provoked state efforts to control industrial practices, considered anathema from the late 1800s to the New Deal. The low point in this jurisprudence was the Supreme Court's 1905 decision in *Lochner v. New York*,[35] which overturned the State of New York's efforts to control the hours that bakers were allowed to work. Despite the dissent's citation of ample evidence demonstrating the severe harm to bakery workers caused by exposure to flour dust, intense heat, and long hours of work in a standing position,[36] the Court refused to even recognize the state law as an exercise of police power, instead dismissing it as a "labor law" that interfered with the constitutional right of freedom to contract.[37]

As the New Deal era unfolded, World War II was fought and won, and America settled into the postindustrialization era, the Supreme Court repudiated *Lochner* in a series of decisions. But these cases stopped mentioning the police power and were instead rationalized as federal deference to the states' prerogatives. For example, *West Coast Hotel Co. v. Parrish*[38] upheld a Washington statute requiring that women be paid a minimum wage, holding that,

> [i]n dealing with the relation of employer and employed, the Legislature has necessarily a wide field of discretion in order that there may be suitable protection of health and safety, and that peace and good order may be promoted through regulations designed to insure wholesome conditions of work and freedom from oppression.[39]

Similarly, in *Day-Brite Lighting v. Missouri*, the Court upheld a Missouri statute allowing workers to be absent from their places of employment for four hours between the opening and closing of election polls:

> Our recent decisions make plain that we do not sit as a super-legislature to weigh the wisdom of legislation nor to decide whether the policy which it expresses offends the public welfare.... [T]he state legislatures have constitutional authority to experiment with new techniques; they are entitled to their own standard of the public welfare; they may within extremely broad limits control practices in the business-labor field, so long as specific constitutional prohibitions are not violated and so long as conflicts with valid and controlling federal laws are avoided.[40]

In a series of articles notable as much for their careful study of history as for their insightful constitutional analysis, Professor Parmet argues that this reluctance to expand the concept of a police power to include widespread and chronic threats to public health, especially where workers were involved, served to deconstitutionalize public health law. The phrase *salus populi supreme lex* was dropped from judicial lexicon, and the Supreme Court "abandoned the quest for the boundaries between the public and private spheres of authority."[41] No one lamented this subtle shift at the time, in large measure because state and federal legislatures were given ample running room by the courts to enact all manner of public health programs, from wage and hour laws to occupational safety laws to environmental regulations.

Professor Parmet concludes that, while it may be difficult to document the tangible effects of deconstitutionalization, the cumulative effect of these decisions is to take the focus off government's legitimate and affirmative responsibility to preserve civil society by protecting public health. The states' decisions to spend a great deal of money and control large swaths of industrial activity are rationalized instead on the basis that they are supported by the majority of the voters and should be scrutinized by the courts only where institutional rights are clearly jeopardized or newly energized principles of federalism are compromised. This commentary could as easily be applied to environmental law. Anchoring the constitutionality of environmental statutes on the Commerce Clause deflects consideration of what government should be expected to do for the people in an affirmative sense. As in the public health context, judicial and legislative debates over the wisdom of environmental policy revolve around the merits and downsides of the individual policy in economic terms rather than the principle that communities should expect government to preserve essential natural resources without which a healthy life is impossible.

Like every scholar firmly grounded in the implications for future events of her historically based theories, Professor Parmet notes the shocking implications of the tragedies that began on September 11, 2001, for public health constitutional doctrines.[42] She suggests that these events should have made clear to every observer that the federal government has a crucial role to play in protecting public health from bioterrorism and such emerging threats as pandemics. In this new and fearful era, all three branches of government have an unavoidable stake in removing the barriers between state and federal responsibilities to protect public health. Similarly, the advent of climate change demands a more flexible interpretation of constitutional intent than the Supreme Court has yet realized.

B. The Federal Police Power

One of the most bizarre and troubling by-products of the September 11, 2001, attacks was the mailing of anthrax to the offices of Senator Thomas Daschle (D-ND) and the NBC News anchor Tom Brokaw.[43] The Federal Bureau of Investigation later discovered that this act of terrorism was almost certainly conducted by a mentally ill American who worked as a senior researcher in a military laboratory at Fort Detrick in Frederick, Maryland.[44] The incident cast in sharp relief the extraordinarily high expectations that Americans harbor toward the national government's ability to protect people from such threats. Had the anthrax spread, the federal government would have organized a coordinated response. Federal officials would have depended heavily on local hospitals, police departments, and emergency medical personnel, but they would have been the ones held accountable for curtailing the spread of the dreadful disease. Indeed, most people – especially the federal officials rushing to sit in the hot seat of such crises (former Pennsylvania Governor Tom Ridge, the newly appointed chief of homeland security for then president George W. Bush, dominated the airwaves in the wake of the crisis) – would have considered anyone who questioned this preeminent role unpatriotic.

Despite this clear manifestation of national consensus on the federal government's obligation to exercise what the law has always described as police powers, conventional theory is that only the states have this authority. As the anthrax episode illustrates, the global challenges of the twenty-first century make this stubbornly constricted reading of the Constitution very much against our national interests. No participant in the national policy debate would ever challenge the proposition that the federal government must play the dominant role in preventing and responding to global threats. And, of course, the federal government has erected an elaborate bureaucratic infrastructure to exercise its police powers (see, e.g., the National Institutes of Health, the Department of Health and Human Services, and the Centers for Disease Control and Prevention).

Similarly, the next generation of public health and environmental laws will be compelled to deal with global threats. The national government has long dominated domestic regulation through a cooperative federalism system that defines the terms and conditions of environmental protection through federal statutes but allows states to volunteer for the responsibility of implementing those rules. Among the strongest principles embodied in that system is that federal standards set the floor – as opposed to the ceiling – of protection and that states can go further if they deem additional protections to be necessary. One statute, the Clean Air Act, even goes so far as to allow California to impose more stringent rules on motor vehicles sold in interstate commerce because the state has acute air-quality problems. But this principle is beginning to unravel in the climate-change context, with many large fossil fuel producers and users demanding preemption of state authority to curb greenhouse gases.[45] Ironically, the most prominent argument these parties make is that the federal government has superior ability to combat this global crisis, which threatens both public health and natural resources.[46] Although I do not agree with the next step of this argument – that states should be excluded from governmental efforts to control climate change[47] – its proponents, among the largest companies in the world, clearly have acknowledged that much more than the free flow of interstate commerce is at stake. Instead, federal control is viewed as crucial in ensuring that the nation makes progress on this global threat.

C. Taxing and Spending

Pouring the hopes for the future of environmental law into the as-yet-unused chalice of the General Welfare Clause is likely to result in overflow, or at the very least a tight fit, unless one specific doctrinal problem is addressed. The Supreme Court held in *United States v. Butler* that the power to provide for the general welfare is directly tied to congressional taxing and spending authority.[48] The 1936 case involved agricultural subsidies designed to control crop prices. The Supreme Court struck down the program because it addressed "agricultural production," a "purely local" industry that only the states were empowered to address.[49] This aspect of the case is no longer good law given the dramatic expansion of federal Commerce Clause authority discussed earlier.[50] More significantly, however, the Court also held that the national government's power to provide for the general welfare is conditioned on its simultaneous exercise of the power to tax to and spend, and this rule remains good law.[51] Most significantly, the Court stated that the "the power of Congress to authorize expenditures is not limited by the direct grants of legislative power found in the Constitution."[52] Accordingly, if Congress is willing to put its money where its mouth is, whether or not it has the power to legislate under the Commerce Clause, it is free to create programs that promote its perception of what is needed to provide for the general welfare.

The federal government spends considerable sums each year to manage federal lands, deploying park rangers and firefighting teams, maintaining the national parks, securing the borders from private-sector incursions, building roads, supervising the preservation of wildlife, and so on. These substantial financial commitments are probably sufficient to satisfy the *Butler* test as applied to the core requirements of the Legacy Act, which would mandate planning to ensure the long-term sustainability of natural resources located on federally owned lands. However, as discussed at the outset, the Legacy Act would also require that proposed action be abandoned or modified if analysis showed that it would threaten long-term sustainability of federally owned natural resources. The power to block such actions should reasonably extend to actions that take place on privately owned land if they would have comparable effects. Opponents of the Act could argue that, unless the federal government subsidized the costs of these consequences, the new law should be read as purely regulatory with respect to private conduct and therefore not involving the exercise of taxing or spending authority. Any number of other legislative formulations that seek to protect future generations could raise comparable questions.

Article I, section 8, concludes that Congress has the power

> To make all Laws which shall be necessary and proper for carrying into Execution the foregoing Powers, and all other Powers vested by this Constitution in the Government of the United States, or in any Department or Officer thereof.

The courts have given Congress wide latitude in designing taxes and spending (e.g., state grants-in-aid) to include prescriptions that affect behavior of either the taxpayer or the recipient of federal largesse.[53] How far they would defer in cases where spending is accompanied by regulation that affects the conduct of private parties is obviously an issue that must be resolved if and when we make the shift in constitutional doctrine. Supporters of the Legacy Act and similar next-generation proposals would urge the most generous leeway, pointing out the very large sums the federal government is spending – and the even larger sums it would be required to spend – if these prohibited actions continued.

Putting the arguments together, then, the strong advantages of grounding the protection of public health and natural resources in the concept of the government's affirmative responsibilities to safeguard the quality of life in a civil society would best be served by recognition of a federal police power anchored on the General Welfare Clause. There remains the question of whether this power, or authority, can traverse the long distance to becoming a judicial enforceable right. For the foreseeable future, as explained in the next section, I think it cannot. The proposition that the Constitution grants Congress authority to provide for the general welfare of the people is not based on any notion that the people, or nature for that matter, have a judicially enforceable right to such protection. If the people are not satisfied with how Congress carries out this mission, their remedy is at the ballot box, not in the courts.

V. The Affirmative Constitution

A. Negative versus Affirmative Rights

Midway through the Reagan administration, long-simmering tensions among constitutional experts erupted into public view. The conflict, which is ongoing, pits conservatives who view the Constitution as primarily important for the "negative" rights it affords individuals against liberals who read "affirmative" rights into the text.[54] A full exposition of this extensive debate is beyond the scope of this chapter. Nevertheless, locating my admittedly ambitious theory on the progressive end of this evolving constitutional scholarship should help readers evaluate it further.

The debate began in 1985 when then attorney general Edwin Meese gave a speech to the American Bar Association articulating the theory of originalism, an approach to interpreting the Constitution that views it as an immutable, transcendent law that is not subject to evolving, arguably inconsistent interpretations.[55] Unless a proponent of a constitutional theory can provide substantial evidence that the framers of the document intended for it to be read the way they think we should read it today, the theory is rejected. Judges should not substitute their personal biases or policy choices for a careful study of what the Constitution's framers had in mind.

Conservative commentators on the bench and in academia have embraced the doctrine, and it has become a central tenet for at least four Supreme Court justices (Alito, Roberts, Scalia, and Thomas), as well as countless judges in the lower courts. Originalism has prompted conservatives – and liberals seeking to persuade conservative judges and policy makers – to search through the documents contemporary to the Constitution, especially the *Federalist Papers*, as well as history books to support claims that the framers in their wisdom intended the result they advocate. Professor Parmet's careful exposition of the framers' awareness of Massachusetts vaccination laws, discussed earlier, is an example of this kind of analysis.[56]

Supreme Court Justice William Brennan responded to General Meese in a speech at Georgetown University a few months later:

> We current Justices read the Constitution in the only way that we can: as Twentieth Century Americans. We look to the history of the time of framing and to the intervening history of interpretation. But the ultimate question must be, what do the words of the text mean in our time. For the genius of the Constitution rests not in static meaning it might have had in a world that is dead and gone but in the adaptability of its great principles to cope with current problems and current needs.[57]

His views have been reiterated by progressive or liberal academics and judges, perhaps most notably by Justice Stephen Breyer in his 2005 book *Active Liberty: Interpreting Our Democratic Constitution*.[58] Under this alternative view, the Constitution establishes a framework for the most successful system of government in human history, but its statement of generalized principles must be interpreted in a flexible way to resolve challenges that the framers could not have anticipated. The most important guidance for judges is their informed consideration of the shifting values shared by the American public, along with the consistent commitment to the protection of minority rights that motivated the framers to establish a constitutional republic rather than a direct democracy.

Not surprisingly, because these discussions have revolved around judicial interpretations of the First, Second, Fourth, Fifth, and Fourteenth amendments, they have focused on government's obligations to avoid interfering with individual autonomy (or negative rights), as opposed to its responsibility to provide benefits for the people (or affirmative rights). For example, challenges to the exercise of state police powers are often brought by individuals placed under quarantine or compelled to undergo vaccination.[59] Judicial efforts to balance the needs of the community against the Bill of Rights did not substantially diminish the notion that the states had substantial discretion to do what was necessary to protect the general public.[60] A smattering of scholars have argued for the recognition of affirmative rights to fundamental needs like health care but have encountered heavy resistance from the academic establishment and the courts.[61]

The landmark case is *DeShaney v. Winnebago County Department of Social Services*,[62] which concerned the tragic story of four-year-old Joshua

DeShaney, who was so severely beaten by his father that he became profoundly retarded. The lawsuit was brought by his mother, who asserted that county authorities had twice returned Joshua to the custody of his father despite their awareness that Joshua was victim of repeated, severe physical abuse in his father's house. These actions violated Joshua's rights under the Fourteenth Amendment and entitled him to recovery under 42 U.S.C. § 1983, the civil rights statute granting a cause of action against government officials who act outside the scope of their authority. Lower courts had rejected these claims. However, the Third Circuit held in a similar case that, once a state or local government learns of abuse and undertakes to protect a child from such danger, it forms a special relationship that imposes an "affirmative constitutional duty" to provide adequate protection.[63] Six state attorneys general, the National Association of Counties, and the National School Boards Association filed amicus briefs in the *DeShaney* case, warning of the floodgates that would open if the Court put a foot wrong by suggesting that the states assumed an obligation to keep people like Joshua safe whenever the social service systems produced decisions about their lives. The Supreme Court granted certiorari to address this conflict in the circuits.

Acknowledging that the facts of the case elicited "natural sympathy," the majority noted that "before yielding to that impulse, it is well to remember once again that the harm was inflicted not by the State of Wisconsin, but by Joshua's father."[64] Because the county had no "constitutional duty" to protect Joshua, its failure to do so, "although calamitous in hindsight," is not a violation of the Due Process Clause.[65] Justice Brennan, writing in dissent for Justices Marshall and Blackmun, significantly did not challenge the majority's view that "the Due Process Clause as construed by our prior cases creates no right to basic government services."[66] But, he added, Joshua's case did not present that question. Rather, Joshua suffered grave injury after the county had already taken action on more than one occasion to place the small child under the control of his father, thereby subjecting him to the possibility of abuse. If their actions were arbitrary, then Joshua and his mother should recover, and the case should be remanded to the trial court for examination of that issue. Given the extreme circumstances of the case, and the enormous pressure that state and local governments can exert on courts by warning of the unforeseen consequences of creating affirmative rights, it is difficult to imagine that a majority of the Court would reverse this position any time soon.

Conditioning the argument for recognition of a federal police power on the acknowledgment that the Constitution grants Congress authority to take action but does not confer on individual citizens a judicially enforceable right to such protections has the great advantage of neutralizing a central assertion of the originalists. It would be a bad thing if unelected federal judges undertook the difficult job of deciding when and how to deploy the government's limited resources to combat such extensive threats. But I advocate that these changes be undertaken by Congress and the executive branch, not the judiciary. As Professor Robin West has argued persuasively in other contexts, the legal academy's preoccupation with the role of judges in making law too often obscures the responsibilities of the other branches to implement the affirmative authority the Constitution grants to Congress.[67]

B. The Aspirational Constitution

If judges are not the sole source of constitutional interpretation and should not control – literally or by implication – how other branches read the Constitution's affirmative grants of authority, how should Congress interpret its responsibilities under the General Welfare Clause? Professor West has argued that the Constitution creates positive obligations to pass laws that will protect citizens against environmental threats.[68] She points to the writings of Thomas Hobbes, John Locke, Thomas Paine, and more recently, John Rawls in defining the nature and scope of those responsibilities. She accepts the *DeShaney* holding as definitive at the same time that she dismisses it as irrelevant to Congress's quest to define its affirmative role. Professor West further argues that the American people believe in the idea that Congress has affirmative duties and are far more concerned about its failure to carry out those duties than they are about the government's interference with the individual rights that so preoccupy constitutional scholars:

> The worry increasingly voiced by American citizens, particularly in Katrina's wake, is that our domestic politics and the state that is its product have become too *wan*, not too voracious, even as our foreign policies have become monstrously outsized. Our shrunken state, incapable

> of either preparing for or mounting an adequate response to a hurricane, incapable of repairing deteriorating bridges or crumbling schools, incapable of responding to public health crises or to a dangerously warming climate, seems, to many of our co-citizens, to be in breach of the most basic, fundamental duties central to a sensible construal of virtually any social compact. Thus, where lawyers look at our government and see the "empire of force" of which Weil spoke, in violation of any number of constitutional norms, many of our co-citizens see, at best, sloth – an empire that is failing or willfully refusing to live up to its most basic obligations.[69]

Time will tell whether a Democratic Congress and the Obama administration will respond to these deep-seated perceptions of government inadequacies or whether they will shrink from these challenges in the face of strident charges that protections will cost too much and drown our children in debt. We may think we cannot afford to deal with climate change and sustainability, but our children almost certainly will not be able to afford to confront these problems if we fail.

Conclusion

Despite their implicit commitment to precedent and a stable interpretation of the Constitution, General Meese and other originalists would be compelled to acknowledge that the Supreme Court has made some stunning, 180-degree turns in its history. In 1896, *Plessy v. Ferguson* upheld the separate-but-equal doctrine in the context of railway travel,[70] only to rectify this disgrace fifty-eight years later in *Brown v. Board of Education*.[71] Yet the Constitution that existed in 1896 was the same as the Constitution that existed in 1954. What had changed were the hearts, minds, and perceived social imperatives of the justices appointed to safeguard it. At this moment on Earth, with the planet's future literally hanging in the balance, it may be time for a similar constitutional moment.

If we stick with Commerce Clause analysis, giving economists free rein to forecast future markets in commodities like clean air and clean water, we can justify incremental but significant changes. The problem with these projections is that the economic value of natural resources left in trust for future generations diminishes to zero over time. Consequently, traditional economic analysis militates against preserving environmental quality for future generations. Supporters of proposals to protect natural resources will seldom win a numbers game unless they limit their preservation goals to a severely constrained short term.

If, in contrast, we read the Constitution as embodying additional values beyond preservation of the marketplace, the horizon of change may well be extended beyond where we already see. The National Environmental Legacy Act and similar breakthrough laws, all of which are necessary to avert the worst consequences of climate change, could be based on the principle of preserving the general welfare that was embraced by the framers and that remains central to Americans' understanding of the rule of law today. Had the justices serving on the Supreme Court in the postindustrialization era been less timid, or less focused on shielding the marketplace from government interference, they might well have considered whether the Constitution's text provided additional authority to protect public health, safety, and the environment.

NOTES

1. U.S. CONST., art. IV, § 3.
2. J. Clarence Davies, *Oversight of Next Generation Nanotechnology* 24 (April 2009), *available at* http://www.nanotechproject.org/process/assets/files/7316/pen-18.pdf.
3. U.S. CONST., art. I, § 8. The Constitution further states in art. I, § 8, that "[a]ll legislative powers herein granted shall be vested in a Congress of the United States."
4. WEBSTER'S THIRD NEW INTERNATIONAL DICTIONARY OF THE ENGLISH LANGUAGE s.v. "commerce" (Philip Babcock Gove & Merriam-Webster Editorial Staff eds., 1993).
5. For a description of these developments, *see* ERWIN CHEMERINSKY, CONSTITUTIONAL LAW PRINCIPLES AND POLICIES 254–75 (3d ed. 2006).
6. *See, e.g.*, Sidney A. Shapiro & Christopher H. Schroeder, *Beyond Cost-Benefit Analysis: A Pragmatic Reorientation*, 32 HARV. ENVTL. L. REV. 433 (2008); Douglas A. Kysar, *Discounting... On Stilts*, 74 U. CHI. L. REV. 119 (2007); RENA I. STEINZOR, MOTHER EARTH AND UNCLE SAM: HOW POLLUTION AND HOLLOW GOVERNMENT HURT OUR KIDS (Univ. Texas Press 2007); Frank Ackerman, *The Unbearable Lightness of Regulatory Costs*, 33 FORDHAM URB. L.J. 1071 (2006); David M. Driesen, *Is Cost-Benefit Neutral?* 77 COLO. L. REV. 335 (2006), and *Distributing the Costs of Environmental, Health, and Safety Protections: The Feasibility Principle, Cost-Benefit Analysis, and Regulatory Reform,* 32 B.C. ENVTL. AFF. L. REV. 1 (2005); Robert R.M. Verchick, *The Case against Cost-Benefit Analysis*, 32 ECOLOGY L.Q. 101 (2005); FRANK ACKERMAN & LISA HEINZERLING, PRICELESS: ON KNOWING THE PRICE OF EVERYTHING AND THE VALUE OF NOTHING (New Press 2004); Catherine A. O'Neill, *Mercury, Risk, and Justice*, 34 ENVTL. L. REP. 11,070 (2004); THOMAS O.

McGarity, Sidney A. Shapiro & David Bollier, Sophisticated Sabotage (Envtl. L. Inst. 2004); Amy Sinden, *Cass Sunstein's Cost-Benefit Lite: Economics for Liberals*, 29 Colum. J. Envtl. L. 191 (2004); Thomas O. McGarity, *Professor Sunstein's Fuzzy Math*, 90 Geo. L.J. 2341 (2002); Lisa Heinzerling, *Regulatory Costs of Mythic Proportions*, 107 Yale L.J. 1981 (1998) and *The Rights of Statistical People*, 24 Harv. Envtl. L. Rev. 189 (2000).

7. U.S. Const., art. I, § 8.
8. For an excellent discussion of these developments in the context of environmental protection, *see* Robert V. Percival, *"Greening" the Constitution: Harmonizing Environmental and Constitutional Values*, 32 Envtl. L. 809 (2002).
9. *See, e.g.*, Wickard v. Fillburn, 317 U.S. 111, 127–28 (1942).
10. New York v. United States, 505 U.S. 144 (1992).
11. United States v. Lopez, 514 U.S. 549 (1995).
12. *Id.* at 561–62.
13. *Id.* at 615–31 (Breyer, J., dissenting).
14. I cheerfully made these arguments myself in Rena I. Steinzor, *Unfunded Mandates and the "New (New) Federalism": Devolution, Revolution, or Reform?* 81 Minn. L. Rev. 97, 154–65 (1996).
15. Morrison v. United States, 529 U.S. 598, 700–01 (2000).
16. *Id.* at 613.
17. *Percival, supra* note 9, at 837.
18. Solid Waste Agency of N. Cook County v. U.S. Army Corps of Eng'rs, 531 U.S. 159 (2001). The Fourth Circuit has gone as far as to declare that Commerce Clause authority does not extend to isolated wetlands. United States v. Wilson, 133 F.3d 251 (4th Cir. 1997).
19. *Id.* at 171–73.
20. *Id.* at 174.
21. *Id.* at 193 (emphasis in original).
22. Rapanos v. United States, 547 U.S. 715 (2006).
23. Linda Roeder, *Clean Water Act: Jurisdiction over Wetlands, Infrastructure, Stormwater Runoff Top Agenda This Year*, 40 Envtl. Rep. 24 (2009).
24. 529 U.S. at 700–01 (2000)
25. Nat'l League of Cities v. Usery, 426 U.S. 833 (1976).
26. Garcia v. San Antonio Metro. Transit Auth., 469 U.S. 528 (1985).
27. U.S. EPA, Summary Report to the President: The Presidential Regulatory Reform Initiative by the Environmental Protection Agency, *reprinted in* 1995 Daily Env't Rep. (BNA) 121 (June 23, 1995) ("We have learned that by focusing on results, not on how results are achieved, we can tap the creativity of Americans to devise cleaner, cheaper, smarter ways of protecting the environment.")
28. For an admittedly extreme but not atypical example, consider former congressman Tom DeLay's comparison of the EPA to the Gestapo. Bruce Burkhard, "Year in Review Congress vs. Environment: Environmental Laws Suffer under GOP-Controlled Congress," *Cable News Network*, Dec. 29, 1995, *available at*

http://www.cnn.com/EARTH/9512/congress_enviro/. An audio recording of Congressman DeLay's comments is available through the online version of the article.

29. For a compelling explanation of the history of public health law, *see* Wendy E. Parmet, *From* Slaughter-House *to* Lochner*: The Rise and Fall of the Constitutionalization of Public Health*, 40 Am. J. Legal Hist. 476 (1996) [hereinafter *Rise and Fall*] (focusing on the full sweep of historical developments from colonial times through the New Deal to the present) and Wendy E. Parmet, *Health Care and the Constitution: Public Health and the Role of the State in the Framing Era*, 20 Hastings Const. L.Q. 267 (1992–93) [hereinafter *Health Care and the Constitution*] (considering especially the framers' expectations about the role of the government with respect to public health during colonial times).
30. Slaughter-House Cases, 83 U.S. 36 (1872).
31. U.S. Const., amend. XIV.
32. Slaughter-House Cases, 83 U.S. at 61.
33. *Id.*
34. *See Rise and Fall*, at 493–501.
35. Lochner v. New York, 198 U.S. 45 (1905).
36. *Id.* at 70–71.
37. *Id.* at 57.
38. W. Coast Hotel Co. v. Parrish, 300 U.S. 379 (1937).
39. *Id.* at 392.
40. Day-Brite Lighting v. Missouri, 342 U.S. 421, 423 (1952).
41. *Rise and Fall*, at 502.
42. Wendy E. Parmet, *After September 11: Rethinking Public Health Federalism*, J.L., Med. & Ethics, 201–11 (2002).
43. David Johnston & Alison Mitchell, *A Nation Challenged: The Widening Inquiry; Anthrax Mailed to Senate Is Found to Be Potent Form; Case Tied to Illness at NBC*, N.Y. Times at A1 (Oct. 17, 2001).
44. Scott Shane & Eric Lichtblau, *Scientist's Suicide Linked to Anthrax Inquiry*, N.Y. Times at A1 (Aug. 2, 2008).
45. *See, e.g.,* Green Mountain Chrysler Plymouth Dodge Jeep v. Crombie, 508 F. Supp. 2d 295, 392 (D. Vt. 2007). In the case, the auto industry argued that California's efforts to pass state laws dealing with this global problem should be preempted by the federal government's exclusive constitutional authority to conduct the nation's foreign affairs.
46. *See, e.g.*, Massachusetts v. EPA, 549 U.S. 497 (2007) (the Clean Air Act covers greenhouse gas emissions and the EPA must make a decision as to whether such emissions constitute a sufficient endangerment to require regulation under the statute).
47. *See, e.g.*, Center for Progressive Reform, William Andreen, Robert Glicksman, Nina Mendelson, Rena Steinzor & Shana Jones, *Cooperative Federalism and Climate Change: Why Federal, State, and Local Governments Must Continue to Partner* (2008), *available at* http://www.progressivereform.org/articles/federalismClimateChange.pdf.

48. United States v. Butler, 297 U.S. 1 (1936).
49. *Id.* at 63–64.
50. CHEMERINSKY, *supra* note 5, at 274.
51. *Id.*
52. *Butler*, 297 U.S. at 64–65.
53. *See, generally*, CHEMERINSKY, *supra* note 5, at 275–81.
54. For an incisive summary of this literature, *see* Robin West, *Progressive and Conservative Constitutionalism*, 88 MICH. L. REV. 641 (1989–90). *See also* Robin West, *The Aspirational Constitution*, 88 NW. U. L. REV. 241 (1993–94).
55. Speech by Attorney General Edwin Meese III before the American Bar Association, Washington, D.C., July 9, 1985, *available at* http://www.fed-soc.org/resources/id.49/default.asp.
56. *Health Care and the Constitution.*
57. Justice William J. Brennan Jr., speech to the Text and Teaching Symposium, Georgetown University, Washington, D.C., Oct. 12, 1985, *available at* http://ww.fed-soc.org/resources/id.50/default.asp.
58. STEPHEN BREYER, ACTIVE LIBERTY: INTERPRETING OUR DEMOCRATIC CONSTITUTION (2005).
59. *See, e.g.*, Jacobson v. Massachusetts, 195 U.S. 11 (1905) (upholding mandatory vaccination law).
60. *Id.*
61. For commentary on both sides of the issue, *see* Frank I. Michelman, *In Pursuit of Constitutional Welfare Rights: One View of Rawls' Theory of Justice*, 12 U. PA. L. REV. 962 (1973); Frank B. Cross, *The Error of Positive Rights*, 48 UCLA L. REV. 857 (2001).
62. DeShaney v. Winnebago County Dept. of Soc. Servs., 489 U.S. 189 (1989).
63. *Id.* at 194. The Third Circuit case was Estate of Baily by Oare v. County of York, 768 F.2d 503 (1985).
64. *DeShaney*, 489 U.S. at 202–03.
65. *Id.* at 201.
66. *Id.* at 203–04.
67. Robin West, *Unenumerated Duties*, 9 U. PA. J. CONST. L. 221 (2006).
68. *Id.*
69. Robin West, *Ennobling Politics*, in LAW AND DEMOCRACY IN THE EMPIRE OF FORCE (H. Jefferson Powell & James Boyd White eds., Univ. of Mich. Press 2009), *available at* http://papers.ssrn.com/sol3/papers.cfm?abstract_id=1172204.
70. Plessy v. Ferguson, 163 U.S. 537 (1896).
71. Brown v. Bd. of Educ., Topeka, Kan., 347 U.S. 483 (1954).

Part II ENVIRONMENTAL COMPETITION STATUTE

8 An Environmental Competition Statute

David M. Driesen

THE PREVIOUS PART OF THIS BOOK HAS BEEN CONCERNED primarily with conserving our environmental legacy. Environmental law, however, must function not only as a force for conservation of the good but also as the generator of a better future. Next-generation environmental law embraces both preservation and dynamic and constructive change. Thus, the next generation of environmental law must figure out how to generate the sorts of changes that will create a better environmental future.

The next generation of environmental law should use economic incentives to creatively stimulate innovation in environmental technology. This chapter proposes an Environmental Competition Statute as a means of stimulating movement toward a more sustainable future. Such a statute would authorize those who achieve low emissions to collect the cost of achieving low emissions plus a premium from competitors with higher emissions.

This chapter briefly explains the value of using this mechanism. It then canvasses the problems with the first and second generation of environmental law that an Environmental Competition Statute can help us overcome. A detailed description of an Environmental Competition Statute follows. The chapter then turns to possible objections to the scheme not addressed in the previous material. It closes with a brief conclusion.

Value of an Environmental Competition Statute

We have achieved a number of advances in material welfare because entrepreneurs seek to get rich by developing and introducing innovations.

Examples include the cellular phone, the personal computer, and various uses of the Internet. Innovators' ability to gain market share through productive change is limited only by their imagination and ability to meet potential demand. Unfortunately, the free market rarely encourages innovations improving the environment, because they usually benefit the public as a whole rather than particular consumers paying for favorable environmental changes.[1]

An Environmental Competition Statute has the potential to encourage contests to improve environmental quality comparable to the ongoing competition to realize other sorts of improvements. It aims to allow the capabilities of innovators free rein in improving environmental quality. It makes it possible for anybody reducing pollution to realize a profit from doing so.

The statute also creates risks for those who fail to advance and innovate, comparable to the risks faced by noninnovators in competitive markets for nonenvironmental goods and services. Just as makers of mainframe computers must adapt to the threat posed by personal computers or risk losing market share, those who fail to adopt the latest environmental technology should lose money to faster-moving competition. This statute allows environmental innovators to prosper at the expense of environmental laggards, thereby allowing environmental markets to function like other competitive markets. In short, an Environmental Competition Statute encourages competition to improve the environment.

Problems with the Existing Law

Most existing law allows government officials' timidity to limit our environmental achievements.[2] The law authorizes federal and state officials to limit the amount of pollution facilities can emit. The officials administering these laws usually must take into account the costs our most antiquated industry will face in thinking about mandating environmental change.[3] They rarely, however, actively consider the economic benefits that those with newer technologies might realize from substantial positive environmental change when establishing new standards.[4] As a result, even when modernization would generate new jobs and greatly improve the environment, government regulations only rarely demand significant changes in approach.

Government officials often feel obliged when setting standards for an entire industry to make sure that every company in an industry can meet the standards it sets.[5] Although the law authorizes and sometimes requires regulations based on the achievements of the best performers,[6] government officials tend to avoid aggressive regulation because of the political problems that tough standards would create.[7] Although in the market for consumer goods competition tends to make the best performers the trendsetters, in environmental law, laggards have a big influence on the quality of environmental performance.

This feeling of obligation leads to standards that do not reflect the full capabilities industry possesses to improve environmental performance.[8] Government officials often base their regulations on the technical capabilities of pollution control technology. Government officials often, however, have limited knowledge of industry capabilities to improve environmental performance. As a result, they tend to demand relatively modest improvement based on well-understood technology. This has been the case, to some degree, even under statutory provisions designed to force technology.[9]

The judiciary plays a role in exacerbating this problem, because industry regularly litigates to challenge rules limiting its pollution. Government officials know that courts can block implementation of rules if judges find the rules unreasonable.[10] Although the relevant statutes only authorize reversal of arbitrary and capricious discretionary decisions, courts sometimes give rules a very hard look. Because officials cannot predict precisely how courts will apply the rather vague standards governing judicial review of agency rules, they tend to shy away from stringent requirements unless they have very good information indicating that facilities have known techniques available for meeting them.

Many policy makers associate this problem of government regulation failing to encourage substantial innovation with command-and-control regulation. But this timidity problem also limits the achievements of emissions trading programs. Emissions trading programs require government officials to set limits on the amount of pollution that polluting facilities emit.[11] The emissions trading law then authorizes polluters subject to those limits to avoid them if they purchase equivalent extra reductions from other facilities, which makes it possible to meet bureaucratically chosen limits efficiently. Government officials develop the emission limits with the costs to old established industry of making changes very much in mind. They therefore usually make demands that do not require basic technological changes significantly improving societal welfare. For example, title IV of the Clean Air Act includes a very well designed emissions trading program for sources of sulfur dioxide causing acid rain. This program has produced some of the reductions needed to address the ecological problems acid rain causes, but it has not encouraged substantial movements toward modern renewable energy technologies.[12] Rather, it has encouraged traditional end-of-the-pipe controls (scrubbers) and some modest pollution prevention (low sulfur coal).[13] The acid rain program has not made the purveyors of the most promising innovative environmental technologies rich. So, it has not functioned to produce the kind of wide-open competition that has enriched people with new ideas providing material benefits to consumers.[14]

The same problem of government timidity would limit the efficacy of pollution taxes. Economists support pollution taxes as an efficient environmental protection instrument. If the traditional U.S. antipathy toward taxes ever abated sufficiently to allow a pollution tax law to pass at all, government officials would have to choose the tax rates to apply to pollution. They would probably find it politically difficult to set rates sufficiently high to stimulate significant innovation in environmentally friendly technologies.

Existing law does not provide a continuous incentive to innovate and go beyond compliance.[15] Even in an emissions trading program, once the operators of facilities regulated by the program have met government set pollution limits, by purchasing credits from overcomplying plants or through local reductions, no incentive exists to go further.[16] Because of this limited demand for credits, only a limited incentive exists to overcomply; rational polluters will only produce as many credits as noncomplying facilities need to achieve compliance, not more. The incentive to improve environmental performance lasts only until the compliance deadline comes up. Emissions trading provides no incentives for net reductions beyond those envisioned by government officials, who set caps with limited information about private-sector capacity for innovation.

Proponents of emissions trading often assert that government officials can remedy the lack of incentive for continuous innovation by setting new limits that apply after a compliance deadline expires. But setting new limits can be politically difficult. Industry can avoid cost by opposing fresh limits, and it frequently does so.[17] Because government responses to the pressures it faces are unpredictable, government regulation, whether by emissions trading or conventional approaches, does not provide a secure climate for investment and deployment of innovative environmental technologies, even though it has secured some significant incremental improvement and occasional innovations.[18]

Pending climate change legislation, if enacted, may provide a more secure climate for investment than previous trading programs because of the presence of meaningful long-term targets. But such targets are unlikely to be wholly adequate to address climate change. Even in the rare case when a trading program provides a good climate for long-term investment, an Environmental Competition Statute can usefully supplement and make up for weaknesses in the cap.

A tax program would provide a continuous reduction incentive but only for a limited class of innovation, those with marginal costs less than the marginal tax rate. Taxes would not provide good incentives for important, cutting-edge technologies that would require significant investments putting their marginal costs above marginal tax rates, even if such investments would lower costs and improve environmental quality in the long run.

The idea for an Environmental Competition Statute arises from experience with second-generation economic incentives. These incentives fall into two categories, negative incentives that penalize pollution, such as pollution taxes, and positive incentives that reward pollution reductions, such as subsidies. Environmental law, however, functions most dynamically when negative economic incentives fund positive economic incentives. Governments occasionally enacted or considered such programs during the second generation of environmental law. Thus, New Zealand imposed licensing fees on fishing, a negative economic incentive, and used the revenue from these fees to pay fishermen to retire, a positive economic incentive.[19] France taxed water pollution and used some of the revenue to fund wastewater treatment. The Regional Greenhouse Gas Initiative, an emissions trading program, limiting greenhouse gas emissions from electric utilities in the northeastern United States, features auctioning of emission allowances, and states may use these revenues to fund energy-efficiency improvements.[20] The California legislature considered a program, Drive +, that would impose fees on consumers purchasing energy-inefficient vehicles and give those fees to consumers purchasing energy-efficient vehicles as a rebate.[21] And finally, New Hampshire officials considered an Industry Average Performance System, which would redistribute pollution taxes to low-polluting companies. An Environmental Competition Statute seeks to build on these cutting-edge, second-generation reforms to stimulate increased innovation.

Increased innovation is important, because innovation increases our capacity to address environmental problems over time and can reduce the cost of doing so. Yet economists recognize that markets generally stimulate insufficient innovation. The reason for this is that developers of innovation cannot capture all of the benefits that innovation creates for society. These positive spillovers (benefits that do not generate rents for the innovator) arise because innovations can contribute knowledge that spurs additional innovation by competitors.[22] These observations about markets' limits in spurring innovation apply to the markets in pollution control technology that first-generation performance standards create and to the markets that second-generation emissions trading programs create. The value of innovation and the limits of markets in encouraging it suggest the need for creative measures to stimulate innovation, such as an Environmental Competition Statute.

A Description of an Environmental Competition Statute

This section begins with a description of the statute's basic features and the rationale for them. It then discusses a host of design issues that a legislature creating such a statute would face. In general, these issues are similar to issues that policy makers confront in designing other market-based and traditional regulatory programs.

The Basics

An Environmental Competition Statute would aim to stimulate a race to the top, a competition to develop and deploy environmentally superior technology. To stimulate this race, an Environmental Competition Statute authorizes those producing products or services with low emissions to collect fees from competitors with higher emissions. These fees should be sufficient to fund the full cost of using and developing an environmentally superior approach and should provide a premium beyond that amount.

Thus, the law would have two components. First, it would set out a requirement that a relatively high polluter pay any low-polluting competitor requesting a fee a dollar amount equal to the amount the low-polluting competitor spent to achieve lower emissions than the high polluter. The low polluter could demand this fee from any higher-polluting competitor it chooses. Second, the legislation would set out a premium that the high polluter must pay beyond the low polluter's cost. For example, the law could require that, on demand, any polluter with higher emissions than the competing company making the demand must pay the low polluter the cost it incurred to achieve low emissions plus 10 percent of its abatement costs.

This approach would allow environmental markets to emulate the economic dynamics of highly competitive markets. In such markets, firms innovate to take market share from other firms. When they innovate successfully, they in effect take money from their competitors, as their revenues increase and their competitors' revenues diminish. The Environmental Competition Statute's transfer payment provision creates this same effect for environmental goods.

Absent such a statute, environmental markets do not produce freewheeling competition for market share to fully meet consumer demand for environmental goods. Consumers want environmental benefits, but because these benefits are public goods, consumers cannot purchase them in free markets. Thus, I may want clean air, but I cannot pay anybody to produce it. No one party can provide me with clean air, because dirty air comes about as a result of the actions of multiple actors, all or most of whom must clean up to produce clean air. This public character of environmental goods (and bads) distinguishes them from private goods, like air conditioners, that one can purchase from a single party.

Government regulation serves to stimulate provision of the public good of environmental quality. But it does so through a less dynamic mechanism than competition to seize market share. It creates a demand for a discrete government-mandated environmental improvement, which can, as we have seen, be inadequate and take insufficient advantage of private-sector capacity to produce environmental improvements.

The kind of economic dynamic an Environmental Competition Statute provides is powerful. It uses fear and greed to motivate innovation, combining an opportunity for profit for innovators with a risk of loss for those who fail to innovate as quickly as their competitors. By doing this, it allows environmental law to emulate the most widely admired feature of free markets, their tendency to stimulate technological advances bettering our lives. Free markets in private goods likewise depend on fear and greed to motivate technological advancements. Opportunities for profit and fear of loss stimulate the risk taking that must occur to create significant technological advances.

Absent such a statute, each polluter often must internalize (pay for) the cost of pollution control itself. But it may externalize (pass on to others) the costs of pollution – a degraded environment and serious public health problems. This asymmetry discourages cleanup. An Environmental Competition Statute allows polluters to systematically externalize the costs of pollution control, just as polluters can now externalize pollutions' costs. This cost externalization frees them to employ all their ingenuity to cleanup.

Sound principles support the idea of an Environmental Competition Statute. In confronting environmental problems, we should "Do the Best We Can."[23] Too often, however, we settle for mediocre environmental standards, standards that demand some improvement but not nearly as much as the market is capable of delivering. This statute tends to foster technological progress by letting the cutting-edge innovators set the pace. Just as in a market for consumer goods and services, a firm must keep up with what the best firms are doing or lose money, this statute likewise requires firms to match the achievements of their best environmental competitors or risk financial consequences.This statute allows firms to profit from environmental leadership and encourages them to truly do the best they can in advancing environmental quality.

An Environmental Competition Statute also helps overcome problems inherent in the economic dynamics of regulation. Frequently, firms resist regulation en masse, and all regulated firms share an interest in defeating enforcement. Because we all finance firms' antienvironmental litigation and lobbying when we purchase the goods they make, they have a lot of money to use in thwarting progress. The Environmental Competition Statute should make about half of the polluting firms into enforcers of the statute. It promises distinct economic benefits to the cleanest firms, which may lead some firms to support such a statute. In these ways, the Environmental Competition Statute seeks to overcome the economic dynamics at the heart of regulatory failure to keep pace with environmentally destructive activities.[24]

Design Issues

Scope

The legislature may make the obligation to pay low-polluting competitors a general requirement for all classes of pollutants and industries or may instead focus on a particular industry and set of pollutants of concern. Congress (or a state legislature) could, for example, enact an Environmental Competition Statute to focus on all emitters of carbon dioxide, the most important greenhouse gas causing global warming. The owner of a new solar plant, for example, could collect all of the costs of plant construction from owners of existing power plants with higher emissions plus a premium – a dollar amount written into the legislation to provide a profit margin for each low emitter. Similarly, makers of vehicles with low carbon dioxide emissions could demand that the makers of vehicles with higher emissions pay the additional costs associated with making their vehicles lower emitting.

Design questions will arise about how to define the class of polluters with responsibilities and opportunities under the Environmental Competition Statute, just as they do under other market-based programs and traditional regulation. Thus, for example, a program that focuses only on emissions characteristics built into cars will not allow consumers, who may influence emissions through their driving behavior, to play in this market. Such a choice may be justifiable in all of these contexts, because of the difficulty of monitoring and regulating individual driving behavior. But it means that we should bear in mind that none of these approaches is a panacea that works for all facets of all programs. We should, however, in defining polluters for purposes of environmental competition, try to capture as much of the pollution generated throughout the life cycle as is feasible.

For simplicity's sake, it may be wise to authorize those who clean up to collect fees from any polluter that they choose to target but require that they choose only a single polluter. The cleaner company will likely choose high-polluting and well-capitalized targets, as it will be easy to prove that they are cleaner than a very dirty company, and it will be easiest to collect from a wealthy company. One could use variation in the premium paid to cleaner companies to encourage collection from the dirtiest companies, but this would either require complicated legislative line drawing or very difficult data collection by the polluter trying to recover its cost.

Congress (or a state legislature) could enact the Environmental Competition Statute without amending any existing law. It would be a means of supplementing basic obligations with incentives to go beyond those obligations or of encouraging new efforts where little has been done (e.g., global warming).

The legislation, however, would function best if it addressed some matters of detail. The legislation might define the pollutants and/or industries it applies to. It would be important to define the industry in terms of broad functions (e.g., the personal vehicle transportation industry), not specific market niches (e.g., sport-utility-vehicle makers). The whole point is to force transfer payments between companies on the basis of environmental performance in meeting basic consumer needs. This requires identification of the bounds of an industry, as only competitors must pay a low-polluting firm under this approach. Because consumers buying cars have a choice between sedans and sport-utility vehicles, for example, defining a category to include all forms of personal transportation makes sense.

This legislation will be most helpful in areas where we anticipate the need for very significant technological change. Climate change is such an area. Scientists suggest that we will need more than a 50-percent cut in global emissions below 1990 levels by 2050 to avoid dangerous climate change.[25] Because developing countries' emissions are expected to rise during most of this period, this may require cuts of 80 percent or more in developed countries' emissions. Because carbon dioxide emissions constitute about 80 percent of the gases on the basis of warming potential, this implies a massive move away from fossil fuels. Such a move will require major technological changes. Other areas may also benefit from such an approach.

Metrics

The Environmental Competition Statute will have to provide some guidance about how to compare the emissions of competing firms. The measurement issue is not fundamentally different from issues in traditional regulation, where we also must figure out how to measure emissions and fairly take into account differences among firms. But in the context of environmental competition, we may profitably treat some of the issues a little differently than we have in other contexts.

An important aspect of the measurement problem involves the choice of a metric on which to base comparisons. A mass-based metric will not work terribly well in this context. Suppose, for example, that one power plant generates one hundred tons of carbon dioxide per year and another generates two hundred tons of carbon dioxide per year. One might think that it would be appropriate to consider the two-hundred-ton facility as the facility with higher emissions and allow the hundred-ton facility to collect fees from the two-hundred-ton facility. This might, however, be inappropriate. Suppose that the two-hundred-ton facility provides electricity to a million people and the hundred-ton facility provides electricity to just one thousand people. It does not seem fair, in such a situation, to consider the larger facility the higher emitter just because it is big and supplies a lot of customers. A better metric would be tons of carbon dioxide per kilowatt hour per year. This would normalize emissions by the amount of pollution per unit of output. In general, this should be the approach. We should measure and compare emissions by the mass of emissions generated annually per unit of output.

This sort of issue arises outside of the Environmental Competition Statute as well. For example, in designing principles for allocating reduction obligations to electric utilities as part of the proposed U.S. cap-and-trade program addressing climate disruption, legislators rejected an allocation formula based on the amount of electricity generated, recognizing that such an approach penalizes utilities that have low per capita emissions because they have funded energy efficiency. Accordingly, some of the bills included energy-efficiency-based adjustments to the electricity-output allocation formula. The per-customer approach seeks to accomplish the same thing in a simpler way.

We could define the unit of output differently, in terms of numbers of customers served in various categories (e.g., residential and industrial). This would prove more complicated to administer fairly but would provide incentives to pay for energy-efficiency improvements, as these would reduce the amount of emissions per customer. These sorts of issues, what metric to choose, are not unique to the Environmental Competition Statute; they also influence the incentives provided by trading programs and traditional regulation. But in all contexts, wise choices of metrics can make incentives more or less powerful and design more or less complicated, as they do when regulators design other market-based programs or traditional performance standards.

Another threshold issue involves deciding whether to focus on emission levels or emission reductions. This issue, too, has its counterparts in existing regulatory programs. Regulators setting a traditional first-generation performance standard can focus on future emission reductions by demanding even percentage reductions from firms, meaning that clean firms must clean up just as much as dirty firms (in percentage terms) to escape liability. Alternatively, they can set absolute uniform emission limits, which require significant reductions in firms with high baseline emissions but much fewer reductions (or none at all) from relatively clean plants.[26] The Environmental Competition Statute likewise could use emission levels as the trigger for liability, authorizing low emitters to collect payments form high emitters. Alternatively, the statute could employ an emissions reduction approach, basing payments on relative amounts of emission reductions after the program was enacted. For reasons that appear below, a properly designed emissions-level approach functions much better than the emission reduction approach.

The emissions-level approach maximizes pressure on dirty plants to clean up. It makes them immediately vulnerable to demands for payment, even without their cleaner competitors undertaking any new projects, because dirty plants will, at the outset, have more emissions than clean ones. If this approach is used, the statute should give plants a few years before any demands for payment can be made to give owners of relatively dirty plants a chance to clean up to escape fee-payment obligations.

An emission reduction approach works less well because it may grandfather in existing emissions. Under this approach, a very dirty coal-fired power plant could reduce emissions and claim a penalty from a natural gas power plant that produced fewer emission reductions, even if the gas-fired power plant has lower emission levels (as gas is inherently cleaner than coal). It minimizes economic dynamic pressure for fundamental technological changes (e.g., fuel choices) and maximizes fairness to existing polluters. It fails to force significant change because it accepts the status quo baseline as a given. Worse, in some contexts, it can reward dirty facilities at the expense of clean competition. For example, under this approach, an existing coal-fired power plant could reduce its emissions slightly and then collect the cost of doing that from a zero-emission solar facility, which cannot reduce its emissions (as it is impossible to go below zero emissions). Where such perverse outcomes are likely, the emission reduction approach should not be used. By contrast, an emissions-level approach maximizes pressures for environmental advances.

Monitoring and Reporting

As with emissions trading, pollution taxes, and performance standards, an Environmental Competition Statute relies on accurate monitoring and reporting of pollution levels. The Environmental Competition Statute, like other economic-incentive-based approaches, will work best in contexts in which reliable monitoring or estimation is feasible. Provisions in the statute should require the use of the best monitoring techniques available. In addition, polluters must report their pollution, not just to the government but also to their competitors. This reporting will make it possible for competitors to compare emissions for the sake of planning environmental improvements to avoid fees and for the sake of deciding who to seek fees from after a low pollution level is achieved. The reporting should take the form of regular postings on an Internet page accessible to all. Because the statute should be based on comparisons of pollution per unit of output level, the reporting should cover both emissions and production numbers.

Definition of Competitors

The Environmental Competition Statute will have to define competitive markets for the sake of establishing who may collect fees from whom. Existing environmental law generally regulates polluters in an industry category, often defined by standard industrial classification (SIC) codes. SIC codes, however, do not comprehensively identify all ideally relevant competitors in a system designed to reward environmentally friendly innovation and apply negative economic incentives to dirtier means of meeting the same consumer goal. In some cases, SIC codes will be too narrow and in some cases too broad. Ideally, someone who develops, for example, a system of integrated pest management (IPM) that makes it possible to increase crop yields with little or no pesticide use should be able to collect a payment from pesticide manufacturers that compete with him or her to increase crop yields. Even if the IPM developers operate a research farm and the pesticide manufacturer operates a pesticide plant, the statute should regard them as competitors (or allow courts to develop a common law of competition based on broad principles).

The application of the statute to a well-defined group of polluters and pollutants with very clear specific definitions of competitors would minimize disputes about who is a competitor. But broader definitions of competitors would produce much more innovation and fundamental change in how we deliver goods and services to consumers.

Discouraging Collusion

The legislation should forbid communication among competitors about how firms plan to respond to the Environmental Competition Statute. Otherwise, they might agree to do nothing, thereby eliminating the incentives to compete. Violation of the provisions should carry very heavy penalties, including jail terms for individuals who commit deliberate violations. Such communication should be regarded as proof of a conspiracy to prevent environmental competition in violation of antitrust principles. Absent such conspiracies, some companies with advanced environmental capabilities will likely seize the opportunity to extract payments from competitors, thereby starting the race to the top. Firms who do not view themselves as environmentally advanced may start beefing up their emission-reducing activities out of fear of becoming a target.

Minimizing Litigation

The legislation should also seek to minimize litigation by providing a dispute settlement mechanism, perhaps through mandatory arbitration. Disputes may arise as to who is a competitor and who has the lowest emissions. Those using continuous monitoring should be presumed to have lower emissions than competitors, unless the competitor can prove otherwise. This will encourage reliable monitoring. Still, legitimate disputes about how to estimate or measure emissions may arise. So, it is desirable to see to it that these quarrels do not become so time consuming as to blunt the program's effects. On the other hand, actions to reduce pollution to get transfer payments or to avoid having to become a payer of one can prove productive even if final settlement is delayed.

The Environmental Competition Statute will not generate complicated environmentally fruitless disputes. The Comprehensive Environmental

Response, Compensation, and Liability Act of 1980, otherwise known as Superfund, has a reputation for generating vexing disputes. This U.S. federal law makes a variety of parties associated with a toxic waste site responsible for that site's cleanup. This has often led to protracted disputes about how to apportion liability among potentially responsible parties.

Superfund, however, has been a notable success in encouraging parties not to create new toxic waste dumps since its enactment in 1980.[27] The Environmental Competition Statute would likely stimulate a comparable scramble to avoid liability. The principle causes of protracted disputes and high transaction costs under Superfund would not exist under the Environmental Competition Statute. Allocating responsibility under Superfund has proved difficult because obtaining good information about the past history of toxic waste dumps (e.g., who dumped, who allowed dumping) has proved difficult and the program creates great uncertainty about the eventual cleanup's scope. By contrast, the Environmental Competition Statute will apply to facilities where the responsibility for pollution clearly belongs with the owner of the facility. It usually will not prove difficult to determine pollution levels, because pollution is ongoing, not past, and liability will arise only after cleanup is completed and documented and the costs completely known. Furthermore, one can structure the Environmental Competition Statute to limit the parties involved to as few as two – one defendant and one plaintiff, thus avoiding the multiparty litigation that has bedeviled the Superfund program.

Concerns That Such a Statute May Raise

Competition offers great prospects for gains and advancements. But it also involves change. And change can excite fear. This section addresses some of the concerns that the Environmental Competition Statute may bring to the fore.

Jobs

Although an Environmental Competition Statute may increase jobs in companies employing new low-emission approaches, it can conceivably cause job losses and even bankruptcy in high-pollution companies. In other areas of life, we accept occasional job losses as the price to pay for improvement.

Hence, nobody argued that we should throttle the personal computer to stave off job losses in the typewriter industry. If we accept these sorts of consequences as the price of progress in delivering better consumer goods or services, we should accept them, when necessary, as a sometimes-necessary cost of environmental progress.

Congress (or a state legislature) could, however, seek to protect workers from some of competition's potential consequences, just as Congress has protected workers from the consequences of other market-based environmental measures. When Congress enacted the acid-rain trading program, it recognized that the flexibility the program offered electric utilities would probably lead to more use of low-sulfur coal. Although this was good for miners in regions producing low-sulfur coal, it was not good for miners in regions producing high-sulfur coal. Decreased demand for high-sulfur coal could lead to job losses in the regions producing it. Accordingly, Congress provided transitional assistance to high-sulfur coal miners when it passed the acid rain program.

If Congress wishes to protect workers from the consequences of competition, however, the legislation providing this protection should reach all form of competition, not just environmental competition. If we wish to have a more humane policy with respect to the disruptions a competitive economy gives rise to in people's lives, it should be a broad form of protection that helps workers hurt by all sorts of market change, not just change produced by environmental laws creating competitive market dynamics.

Risk-Risk Trade-offs

When an Environmental Competition Statute targets one form of pollution, those reducing or eliminating the target pollutant may respond with measures that create different risks from those which the statute targeted. This problem is not unique to Environmental Competition Statutes; it arises under first- and second-generation programs as well. Still, regulators should anticipate problems that might arise under such a statute. For example, if they do not wish to encourage payments from coal-fired power plant operators to nuclear power plants, because of the risks involved in nuclear power, they should draft provisions prohibiting that. Unanticipated problems, however, can arise in any program that affords industry technological choices.[28]

Cost

The Environmental Competition Statute prototype I have outlined lacks a clear cost constraint. Under the pure form of this approach sketched earlier, clean producers can collect the cost of their cleanliness from dirty competitors no matter how costly the clean approach happened to be. A lack of cost constraint may be useful when addressing extremely serious problems that require substantial innovation, like global warming.

In practice, however, such a statute would not produce entirely unconstrained costs.[29] Producers seeking to introduce cleaner processes must make sure that those processes are not so expensive to operate as to bankrupt them. If they go bankrupt, they are not a competitor that can claim compensation for cleanup. They also must spend money before they collect it and some risk exists that their competition may cleanup as well, so there remains some risk in spending too much without realizing sufficient improvements to collect from a competitor with some financial capability to make the required payments. Even though these economic constraints will apply in practice, the statute will still leave opportunities for those confident that they can beat their competitors' environmental performance without insane expenditures.

Additional cost constraints would limit the statute's effectiveness but still leave scope for significant improvements. The best way to provide an additional cost constraint would be to make after the fact adjustments if costs prove excessive. An ex post approach to cost adjustment would make the program respond to actual costs rather than cost projections, which often prove inaccurate. This constitutes a substantial advantage. The adjustments could include suspending the program, putting a price cap on transfer payments, or limiting the premium paid above the cost of pollution control. All of these measures, however, would compromise the program's environmental effectiveness.

Geographic Scope

A jurisdiction enacting an Environmental Competition Statute must also decide how to address emissions generated by activities outside the jurisdiction enacting the law. This concern arises because most markets feature competition across geographic boundaries. These issues are complicated enough that identifying an industry with substantial competition solely within a jurisdiction as a target for early experiments with the Environmental Competition Statute concept commends itself as a strategy. Under this approach, only facilities in the regulating jurisdiction could collect fees, and only facilities with that jurisdiction would have to pay. We will see, however, that it is possible to handle interjurisdictional competition more robustly.

Before describing the interjurisdictional issues and ways to address them, it is worth noting that traditional environmental law, including emissions trading, faces similar issues.[30] Thus, for example, when the EPA required the reformulation of gasoline to reduce air emissions from cars, effectiveness required that refiners outside the United States also reformulate their gasoline.[31] Otherwise, gasoline from foreign refiners sold in the United States would undermine the program's effectiveness. Accordingly, the United States required foreign refiners to comply.[32] Similarly, traditional regulation has to address transboundary impacts of production facilities' direct pollution, and it usually has done so by some process of agreements among jurisdictions. Yet we shall see that these old issues take a slightly different shape in the context of an Environmental Competition Statute.

It seems clear that a government has jurisdiction to demand that polluters with its territory pay polluters with lower emissions, as required by an Environmental Competition Statute. This jurisdiction would suffice to justify demanding that polluters within the jurisdiction enacting an Environmental Competition Statute pay polluters outside the jurisdiction with lower emissions as well as polluters within it. But the question of whether a jurisdiction can demand payments from polluters outside its jurisdiction to polluters within the jurisdiction may prove more complex. For example, assume that a petroleum refinery in California produces carbon dioxide emissions. It competes with refineries in Texas to sell oil on the interstate market. California might want to force its polluters to compete to reduce refinery carbon dioxide emissions. This would require that California law allow Texas refiners to collect fees from California refiners with lower emissions, which is not jurisdictionally problematic, as the collection would be against a California facility under California law. But it is not as clear that California would have regulatory jurisdiction to demand that Texas refiners pay California refiners with lower emissions. The same question could arise on the national level. For example, could the United States demand that Venezuelan refiners pay U.S. refiners with lower emissions under a U.S. Environmental Competition Statute?

As a general matter, states can tax foreign polluters for activities in the state. Thus, California, for example, would have jurisdiction to force a Michigan car company that sells automobiles in California to pay another car company that also sells cars in California for pollution from the cars driven or sold in California.[33] It is possible that California could also regulate a company that contributed emissions that affected California.[34] But this category might include any company in the world that emitted carbon dioxide, so courts might be tempted to limit the reach of such an exercise of regulatory jurisdiction. In the climate-change context, emissions everywhere affect any state's welfare. Outside of that context, a state might have difficulty regulating facilities outside of its jurisdiction that compete with facilities in the jurisdiction but emit nothing that affects the regulating states.[35] Either a state or the federal government would have to consider limiting a program to embrace less than the entire market that its companies compete in under an Environmental Competition Statute. Even with such limitations in place, such programs would spur a great deal of innovation. And Congress possesses the authority to remove impediments to state Environmental Competition Statutes arising from the Dormant Commerce Clause, the source of most of the potential restraints just mentioned.[36]

Furthermore, states could reach informal agreements or create interstate compacts with congressional approval to broaden the reach of their programs. And nations could broaden the reach of their programs through treaties.

Both states and federal governments would have to conform their programs to relevant law encouraging free trade. Nation-states must conform to World Trade Organization (WTO) agreements, such as the General Agreement on Tariffs and Trade (GATT) and the General Agreement on Trade and Services (GATS), and regional trade agreements. Similarly, U.S. states must conform to the U.S. Supreme Court's Dormant Commerce Clause jurisprudence, which infers limits on state regulation and taxation from congressional authority to regulate interstate commerce. Under all of these free-trade legal regimes, polities usually may not discriminate against companies outside their jurisdiction.[37] This means that governments must resist the temptation to make an Environmental Competition Statute a one-way street, absent a very strong justification.[38] If states demand that out-of-state companies with high emissions pay in-state companies with low emissions, they must also demand that in-state companies with high emissions pay out-of-state companies with low emissions. Programs that reach out-of-state polluters must be carefully crafted to avoid adverse rulings under free-trade law and to conform to limits on state regulatory jurisdiction.

Takings

Many countries prohibit the government from taking private property without just compensation. In most places, this poses no problem for an Environmental Competition Statute because the approach does not involve a government taking of private property. The U.S. Supreme Court, however, has created a unique body of law based on the idea that if government regulation goes "too far," it constitutes a taking, thus triggering a government compensation duty. Companies would probably challenge this law as a taking, both in the United States and possibly in Canada and Mexico, under the North American Free Trade Agreement (NAFTA). But this challenge should fail. The Supreme Court has held that laws requiring monetary transfers without requiring transfer of particular property do not implicate the takings clause.[39] The NAFTA tribunals are extremely unpredictable, but they should not go beyond U.S. law on this, as Canada and Mexico have no regulatory takings doctrine and there is no firm support for such a doctrine in the text of the NAFTA agreement.

A challenge on substantive due-process grounds should also fail (a possibility in the United States, at least). The U.S. Supreme Court upholds all laws having a mere rational basis under this doctrine. Seeking to advance environmental protection through competition may be controversial, but it certainly meets the minimal standards for rationality that govern substantive due-process cases.[40]

Although the Court has upheld laws transferring funds from companies to other private parties, it has struck down an especially unfair retroactive application of one such law.[41] Given the changing composition of the Court and the concern the Court has expressed about retroactive legislation, designers of Environmental Competition Statutes might wish to limit the creation of retroactive liability that might appear unfair to the Court. A simple way to do this is to allow three years after the law goes into effect before any liability can apply, which sound design demands anyway. This gives those potentially subject to liability an opportunity to reduce their emissions and thus their liability, and it avoids retroactive liability. After all, the law's purpose is to stimulate emission reductions, not payments. The prospect of payments serves only as a means toward the end of stimulating competition to clean up.

Conclusion

An Environmental Competition Statute can unleash private-sector capacity to improve the environment with little reliance on frequently lethargic government processes. In this sense, it emulates free market dynamics more faithfully than emissions trading, the signature reform of second-generation environmental law. It allows firms exercising environmental leadership to prosper, thereby discouraging laggards from resisting change. It can help usher in a more successful third generation of environmental law.

NOTES

1. *See* David M. Driesen, The Economic Dynamics of Environmental Law 98–102 (2003) (discussing the private market's limitations in encouraging innovation protecting the environment).
2. *See id.* at 112–22 (discussing the structure and economic dynamics of government decision making in detail).
3. *See, e.g.*, Nat'l Renderers Ass'n v. EPA, 541 F.2d 1281, 1288–89 (8th Cir. 1976) (finding a water-pollution rule arbitrary because the EPA did not adequately consider whether costs would affect the economic viability of medium-sized facilities).
4. *Cf.* Miguel Mendonca, Feed-in Tariffs: Accelerating the Deployment of Renewable Nuclear Energy 43 (2007) (Germany's feed-in tariff system to encourage renewable energy created job growth in the renewable energy sector).
5. *See, e.g.*, Nat'l Lime Ass'n v. EPA, 627 F.2d 625, 632 (D.C. Cir. 1980) (reversing a performance standard because the EPA could not adequately show that its limited data adequately took into account operational variables throughout the industry).
6. *See, e.g.*, 42 U.S.C. § 7412(d)(3).
7. *See, e.g.*, Nat'l Lime Ass'n v. EPA, 233 F.3d 625, 632 (D.C. Cir. 2000) (remanding because EPA ignored statutory commands to show that all sources can achieve the standards set under the most adverse conditions).

8. *See generally* Thomas O. McGarity, *Some Thoughts on Deossifying the Rulemaking Process*, 41 DUKE L.J. 1385 (1992).
9. David M. Driesen, *Distributing the Costs of Environmental, Health, and Safety Regulation: The Feasibility Principle, Cost-Benefit Analysis, and Regulatory Reform*, 32 B. C. ENVTL. AFF. L. REV. 1, 14–16 (2005) (explaining how judicial demands for a rational basis for technology-based rules have limited technology forcing). Bruce La Pierre, *Technology-Forcing and Federal Environmental Protection Statutes*, 62 IOWA L. REV. 771, 805–31 (1977) (contending that judicial requirements that agencies identify at least one technology capable of achieving their promulgated standards limited agencies' ability to force technology); cf. Note, *Forcing Technology: The Clean Air Act Experience*, 88 YALE L.J. 1713, 1718–19 (1979) (claiming that state plans did force some technological improvement).
10. *See generally* Frank B. Cross, *Shattering the Fragile Case for Judicial Review of Rulemaking*, 85 VA. L. REV. 1243 (1999).
11. *See* Michael Grubb et al., *Allowance Allocation in the European Emissions Trading Scheme: A Commentary*, 5 CLIMATE POL'Y 127, 127 (2005) (describing the "allocation of allowances" as "the most... important step" for "any emissions trading system"); David M. Driesen, *Is Emissions Trading an Economic Incentive Program? Beyond the Economic Incentive/Command and Control Dichotomy*, 55 WASH. & LEE L. REV. 289, 324 (1998); Tex. Mun. Power Agency v. EPA, 89 F.3d 858, 861 (D.C. Cir. 1996) (describing the need for limits as a "necessary aspect" of "any" emissions trading program).
12. *See* A. DENNY ELLERMAN ET AL., MARKETS FOR CLEAN AIR: THE U.S. ACID RAIN PROGRAM 130 (2000).
13. *See* David M. Driesen, *Does Emissions Trading Encourage Innovation?* 33 ENVT'L L. REP. 10,094, 10,105 (2003); Byron Swift, *Command without Control: Why Cap-and-Trade Should Replace Rate Standards for Regional Pollutants,* 31 ENVT'L L. REP. 10,330, 10,332 (2001) (describing scrubbers and low-sulfur coal as the principal compliance techniques).
14. *Cf.* Margaret R. Taylor, Edward S. Rubin & David A. Hounshell, *Regulation as the Mother of Invention: The Case of SO_2 Control*, 27 LAW & POL'Y 348, 370 (2005) (finding less innovation under the acid rain program than under the command-and-control regime preceding it); David Popp, *Pollution Control Innovations and the Clean Air Act of 1990*, 22 J. POL'Y ANAL. & MGMT. 641 (2003) (finding more patenting of environmental technology under command and control than under the acid rain trading program, but finding a different type of innovation under trading).
15. *See* Driesen, *supra* note 3, at 10,099–101 (explaining in detail why a trading program fails to provide continuous incentives for environmental improvement); Driesen, *supra* note 9, at 324–27 (same).
16. *See* Robert W. Hahn & Robert N. Stavins, *Incentive-Based Environmental Regulation: A New Era from an Old Idea*, 18 ECOLOGY L.Q. 1, 8–9 & n. 33 (1991) (recognizing that emissions trading tends to reach an equilibrium).

17. *Accord* Andrew McFee Thompson, Comment, *Free Market Environmentalism and the Common Law: Confusion, Nostalgia, and Inconsistency,* 45 Emory L.J. 1329, 1359 (1996) (noting the pressures that bureaucrats face to overallocate allowances in a trading scheme); *see, e.g.*, Inho Choi, *Global Climate Change and the Use of Economic Approaches: The Ideal Design Features of Domestic Greenhouse Gas Emissions Trading and an Analysis of the European Union's CO_2 Emissions Trading Directive and the Climate Stewardship Act*, 45 Nat. Resources J. 865, 902–03 (2005) (describing California's RECLAIM program as a failure because caps were set too high); Axel Michaelowa & Sonja Butzengeiger, *EU Emissions Trading: Navigating Between Scylla and Charybdis*, 5 Climate Pol'y 1, 5 (2005) (explaining how lobbying in the European Union led to goals in the first phase of its emissions trading scheme that provided for little departure from business-as-usual levels of carbon emissions); Grubb et al., *supra* note 11, at 132–33 (same); Tex. Mun. Power Agency v. EPA, 89 F.3d 858, 861 (D.C. Cir. 1996) (involving a claim to additional emission allowances); Indianapolis Power & Light Co. v. EPA, 58 F.3d 643, 647 (D.C. Cir. 1995) (same); Madison Gas & Elec. Co. 25 F.3d 526, 526 (7th Cir. 1994) (same); Mononganhela Power Co. v. Reilly, 980 F.3d 272, 272–74 (4th Cir. 1992).
18. *See* Driesen, *supra* note 13, at 10, 103–05.
19. *See* Tom Tietenberg, *Using Economic Incentives to Maintain our Environment*, 33 Challenge 42, 43 (1990).
20. *See* David M. Driesen, *The Changing Climate for United States Law*, 1 Climate Change L. Rev. 33, 38 (2007) (discussing movement toward auctioning under the Regional Greenhouse Gas Initiative).
21. Nathaniel Greene & Venessa Ward, *Getting the Sticker Price Right: Incentives for Cleaner, More Efficient Vehicles,* 12 Pace Envt'l L. Rev. 289, 346 (1998).
22. *See* Gregory N. Mandel, *Promoting Environmental Innovation and Intellectual Property Innovation: A New Basis for Patent Rewards*, 24 Temple J. Envtl. L. & Tech. 51, 56 (2006) (if a person "builds a better mousetrap," others may copy it); Richard A. Posner, Catastrophe: Risk and Response 123–24 (2004) (third parties' ability to use information makes it difficult for inventors to keep all the value their inventions create). *See generally* Brett Frischman & Mark A. Lemley, *Spillovers*, 107 Colum. L. Rev. 257 (2007).
23. *See* A New Progressive Agenda for Public Health and the Environment: A Project of the Center for Progressive Regulation 57–70 (Christopher H. Schroeder & Rena Steinzor eds. 2004) (discussing this concept as a principal to guide environmental law).
24. *See* Driesen, *supra* note 1, at 113–35 (analyzing the dynamics of this failure in detail). *Cf.* David Goldstein, Saving Energy Growing Jobs: How Environmental Protection Promotes Economic Growth, Profitability, Innovation, and Competition, 172–76 (2007) (explaining how trade associations repress competition to profit from environmental protection and pressure the government to adopt weak standards or none at all).

25. *See* James E. Hansen, *A Slippery Slope: How Much Global Warming Constitutes "Dangerous" Anthropogenic Interference*, 68 CLIMATE CHANGE 269, 277 (2005) (stating that a two-degree Celsius temperature rise "almost surely takes us well into the realm of dangerous" climate change); Malte Meinshausen, *What Does a 2°C Target Mean for Greenhouse Gas Concentrations? A Brief Analysis Based on Multi-Gas Emission Pathways and Several Climate Sensitivity Uncertainty Estimates*, in AVOIDING DANGEROUS CLIMATE CHANGE 269–70 (Hans Joachim Schellnhuber et al. eds. 2006) (estimating that limiting temperature rise to less than two degrees Celsius likely requires a 55 percent reduction below 1990 emission levels by 2050).
26. *See* DRIESEN, *supra* note 1, at 193–95 (discussing the differences between percentage reduction and fixed-level standards).
27. *See* Klaus Lindegaard, *Environmental Law, Environmental Globalization, and Sustainable Techno-Economic Evolution,* in ENVIRONMENT, TECHNOLOGY, AND ECONOMIC GROWTH 141 (Andrew Tylecote & Jan van Der Straaten eds. 1998) (reporting a 51.8 percent reduction in waste generation between 1981 and 1985).
28. *See* David M. Driesen & Amy Sinden, *The Missing Instrument: Dirty Input Limits*, 22 HARV. ENVTL. L. REV. 65 (2009).
29. *See* DRIESEN, *supra* note 1, at 158.
30. *See, e.g.*, Ozone Depleting Chloroflurocarbons, Proposed Production Restriction, 45 FED. REG. 66,727, 66,732–33 (1980) (discussing options for addressing imports of ozone-depleting chemicals under a proposed trading scheme to reduce their domestic production).
31. *See* George E. Warren Corp. v. EPA, 159 F.3d 616, 618–19 (D.C. Cir. 1998) (explaining that antidumping provisions of the Clean Air Act required that both foreign and domestic refiners comply).
32. *See id.* (describing and resolving controversy over compliance methodologies for foreign refineries).
33. The problems underlying this discussion would arise if the Supreme Court applied its Dormant Commerce Clause jurisprudence to an Environmental Competition Statute. Because this statute would not tax – it does not raise revenue for the state – it is not entirely clear that the principles of Dormant Commerce Clause tax jurisprudence would apply to it.
34. *See* Nat'l Geographic Soc'y v. Cal. Bd. of Equalization, 430 U.S. 561 (1997) (requiring only some "minimum connection" between the state and the taxed entity); Walter Hellerstein, *Deconstructing the Debate over State Taxation of Electronic Commerce*, 13 HARV. J.L. & TECH. 549, 552 (2000) (discussing the consensus among tax scholars that states should be able to tax companies having no physical presence in the taxing state).
35. *See* Quill Corp. v. North Dakota, 504 U.S. 298 (1992) (holding that states cannot impose a sales or use tax on a company lacking a physical presence in a state).
36. *See id.* at 318.
37. *See* Healy v. Beer Inst., 491 U.S. 324, 336–37 (1989) (applying strict scrutiny to discriminatory measures).

38. *See* New Energy Co. v. Limbach, 486 U.S. 269 (1987) (invalidating an Ohio tax credit given only to local ethanol producers despite a claim that the credit helped protect the environment).
39. *See* E. Enters. v. Apfel, 524 U.S. 498, 539–47, 554–58 (1998) (concurring and dissenting opinions).
40. *See generally* Collins v. City of Marker Heights, 503 U.S. 115, 125 (1992) (unanimous opinion) (discussing Court's reluctance to expand the substantive-due-process doctrine).
41. *Compare* Pension Benefit Guar. Corp. v. Connolly, 475 U.S. 211 (1986) (upholding requirement that private companies fund retirees' pensions after terminating a retirement plan); Usery v. Elkhorn Mining Co., 428 U.S. 1 (1976) (upholding requirement that mining companies compensate former employees with black lung disease) *with Eastern Enterprises*, 524 U.S. 498 (striking down retroactive liability for black lung disease for a company that never promised health protection from black lung disease).

9 Climate Change, Federalism, and Promoting Technological Change

David E. Adelman

THIS CHAPTER EXAMINES THE LEVEL OF GOVERNMENT – STATE versus federal – at which an Environmental Competition Statute could be most effectively implemented. After years of inaction, the federal government is now debating legislation to address climate change. On May 21, 2009, the House Energy and Commerce Committee took the unprecedented step of voting for the American Clean Energy and Security Act of 2009, which would establish, among other important policies, a comprehensive cap-and-trade program for controlling emissions of greenhouse gases (GHGs).[1] Recent federal action follows a period of rapid policy development by state and local governments. In the absence of strong federal leadership, a growing number of states have filled the void in climate policy with a broad array of programs, including regulation of GHG emissions from vehicles and power plants, renewable energy mandates, GHG emissions registries, and energy-efficiency initiatives.[2]

The question addressed in this chapter is how state initiatives can operate in conjunction with federal programs to induce the technological change needed to mitigate climate change. I show that promoting innovation is a distinct regulatory end that is subject to a market failure – technology spillovers – unrelated to the negative externalities that have traditionally justified environmental regulation.[3] This distinction is significant because critics of state action have questioned whether anything is left for the states

This chapter is a condensed version of an article coauthored with Kirsten Engel, *Reorienting State Climate Change Policies to Induce Technological Change*, 50 ARIZ. L. REV. 851 (2008).

once the federal government acts or even whether state initiatives are per se misguided given the global scale inherent in reducing GHG emissions.[4] Indeed, existing rationales for state climate programs reflect these limits insofar as they view them primarily as a means of prompting federal action or policy innovation.[5]

This chapter challenges the view, endorsed most prominently by the George W. Bush administration, that climate change policies should be set at the international level, leaving only secondary roles for federal or state policies.[6] Economists focused on the principle of fully internalizing the costs and benefits of regulations view federal or limited multinational regulation skeptically, and state policies even more so. Critics also point to countervailing market dynamics. They argue that reductions in GHG emissions will be eroded through emissions leakage, which refers to the circumstances in which GHG reductions in regulated jurisdictions are offset by corresponding increases in emissions from unregulated territories as a result of competitive market conditions. The possibility of adopting an Environmental Competition Statute to address GHG emissions would face these same objections.

None of these arguments refutes the capacity of state (or federal) policies, including an Environmental Competition Statute, to promote technological change. Instead, critiques of state policies have focused almost exclusively on their capacity to reduce GHG emissions, ignoring their capacity to achieve the distinct objective of inducing technological change. Thus, although critics have convincingly demonstrated the degree to which state efforts to reduce GHG emissions are undermined or insignificant, their analyses are incomplete.[7]

This one-sided focus is counterproductive given the importance of new technologies to climate change mitigation. It is also out of step with the economic literature on the capacity of government policies to induce technological change. Economists have noted that "[t]he effect of environmental policies on the development and spread of new technologies may, in the long run, be among the most important determinants of success or failure in environmental protection."[8] Nor is this a naive brand of technological optimism; strong evidence exists that widespread adoption of existing technologies on its own could "solve the carbon and climate problem for the next half century."[9] The current debate over federalism and climate change is therefore not only incomplete, but also overlooks an essential element of climate change policy.

This chapter seeks to refocus attention on the independently valuable objective of inducing technological change through measures like an Environmental Competition Statute. To evaluate the capacity of state and federal legislation to promote technological development, the chapter analyzes the two types of market failure – environmental and technological – relevant to climate change policy. It then explores several avenues for states to induce technological change, whether through technological innovations or adoption of existing technologies. Market dynamics will differ according to whether states adopt policies in the shadow of federal legislation or instead act alone; both scenarios are examined herein. The chapter demonstrates that technology-forcing laws, such as an Environmental Competition Statute, can be effectively implemented by state and federal governments. The chapter concludes with specific recommendations for harmonizing state and federal climate change policies.

Climate Change Market Failures

According to conventional economic theory, environmental regulation is justified when businesses and consumers fail to internalize the negative environmental impacts of their actions. Environmental policy is not limited, however, to deterring or restricting behavior with bad environmental consequences; it also seeks to promote development of new technologies that reduce abatement costs and enable more aggressive action. This latter objective is difficult to meet because it implicates two distinct market failures: traditional negative environmental externalities and the positive externalities associated with uncompensated technology spillovers.

Efforts to regulate climate change are particularly susceptible to these forms of market failure. The negative impacts of GHG-emitting activities are diffused globally, although risks from climate change are by no means distributed uniformly. Similarly, technology spillovers are likely to be large because of the global importance of environmental technologies relevant to mitigating climate change and because of the modest protections for and volatile politics surrounding intellectual property rights in many countries.

The two subsequent sections examine the capacities of federal and state climate change policies to address these externalities. Although state regulation is clearly inferior to national (or coordinated international) regulation in achieving GHG emissions reductions, I show that it has significant potential to complement federal policies and has certain advantages over them in promoting technological change. The third section analyzes how the federal system can accommodate the respective strengths of national and state policies.

Traditional Environmental Externalities

Climate change produces a classic form of pollution externality: industrial producers and consumers alike reap the benefits of activities that generate GHGs while imposing the costs of the emissions on the world as a whole. Because polluters internalize only a fraction of these costs, their incentives to reduce them are only a fraction of what would be economically efficient. Accordingly, traditional economic theory dictates that the global scale of climate change necessitates international regulation.

The dominance of state climate change initiatives in the United States runs contrary to the predictions and principles of classical economic theory. Drawing on this theory, opposition to state programs rests on two primary arguments: (1) state-level programs, by virtue of their small geographic scope, will be too small to have a meaningful impact on global GHG emissions, and (2) state climate regulation is vulnerable to emissions leakage, which involves market responses that can offset emissions reductions in a regulated jurisdiction.

Regulation of GHG emissions from electric power plants illustrates the limited scale of state and regional action. Even when the results of aggressive state programs are aggregated, the volume of emissions reduced constitutes just a small fraction of total emissions in the United States. A useful benchmark here is the group of states committed to or considering involvement in the Regional Greenhouse Gas Initiative (RGGI) – eleven states and the District of Columbia in all.[10] The direct impact of RGGI is minimal because carbon dioxide emissions from power plants (its exclusive focus) in the eight committed states constitute only 2 percent of U.S. emissions from this sector (this rises nominally to 4.5 percent for all eleven states and the District of Columbia).[11] These statistics reveal that state-level standards would have to be twenty times more stringent than a federal standard to achieve a comparable reduction in GHG emissions.

The vulnerability of state programs to emissions leakage weakens the case for state action further. Even if a state program constituted a significant fraction of national GHG emissions, its emissions reductions could be substantially eroded or nullified by emissions leakage to other jurisdictions. Economists have identified three sources of emissions leakage – the first driven by falling fuel prices, the second by reductions in production levels, and the third by relocation of industries to unregulated jurisdictions.

The first two leakage scenarios follow from standard supply-and-demand economics. The logic is simple: insofar as state regulation diminishes demand for carbon-based fuels, it will trigger a drop in global fuel prices (higher local prices or caps would decrease demand in the regulated jurisdiction) that, in turn, will cause offsetting increases in fuel consumption in unregulated areas. The degree to which emissions reductions within a regulated territory are offset will depend on the sensitivity of demand to price in unregulated areas; the more price sensitive it is, the more complete the offsetting emissions will be.

The second scenario is a variation of the first, but it involves reductions in the production of goods within a regulatory territory that are driven by increases in costs of production caused by regulation (i.e., higher energy costs). Similar to the first scenario, a reduction in production levels within a regulated territory will lead to compensating production increases in unregulated areas. Here, reductions in supply lead to increases in product prices, which motivate producers outside a regulated jurisdiction to expand their production levels. Furthermore, if relatively efficient production in a regulated area is replaced by production with higher per-unit GHG emissions, the net result could be greater overall GHG emissions.

The significance of price – and supply-based leakage is dependent on the magnitude of actual reductions and market dynamics. If reductions in carbon-fuel demand or product supply are small, the effects will be nominal unless price or supply is extraordinarily sensitive (elastic) to demand or price, respectively. Thus, because leakage rises with impacts on price or supply, the small scale of state programs makes them less susceptible to price and supply leakage.

The third form of leakage arises when production facilities are relocated outside of a regulated area. For industries that are sensitive to energy prices, differentials in energy prices between regulated and unregulated jurisdictions may justify incurring short-term relocation expenses. Consequently, instead of reducing GHG emissions, a regulating state would merely prompt companies to relocate outside of its borders and continue to emit GHGs at similar (or possibly even greater) levels.

Economic modeling studies support claims that leakage can undermine efforts to regulate GHG emissions at the regional (e.g., all Organisation of Economic Co-operation and Development member countries) or national levels. Estimates using standard economic models predict leakage rates ranging from about 5–20 percent, although some investigators have predicted rates as high as 41 percent.[12] The most dramatic results have involved undifferentiated goods (i.e., fungible commodities) and more complex economic models that take into account the potential for companies to migrate to unregulated jurisdictions. Under these circumstances, leakage rates can range as high as 50–130 percent, implying that state or regional efforts to reduce GHGs emissions can cause a net increase in global emissions.[13]

The potential significance of leakage is therefore highly variable. For industries that produce differentiated goods, the threat of leakage is relatively low, whereas for industries that produce undifferentiated commodities, it can wipe out or even overwhelm GHG emissions reductions from regulatory programs. But for industries that are not mobile, whether because of the nature of the good, unique geographic production needs or human capital constraints, the potential magnitude of leakage is unlikely to offset GHG reductions in a regulated territory.

Not all leakage is negative, though. Technology spillovers can cause it to be positive; that is, regulation in one jurisdiction can precipitate reductions of GHG emissions in unregulated areas. Under this scenario, regulation in a jurisdiction promotes diffusion of new technologies that reduce abatement costs. If technologies, for example, increase energy efficiency and the price sensitivity of carbon-based fuels is sufficiently high, it may be profitable for industries in unregulated areas to adopt them, causing a net reduction in GHG emissions both inside and outside a regulated jurisdiction.[14]

The potential for positive technological spillovers demonstrates the counterbalancing effects of technological change on GHG emission levels. It is also important to recognize that state policy makers have strong incentives to promote the spread of effective technologies and policy experience to other governments precisely because climate change is a global problem. There is already "abundant" evidence of this occurring, with recent studies suggesting that regulatory programs, subsidies, or a combination of the two that induce development and adoption of cost-effective technologies can eliminate leakage.[15] These observations anticipate the important role of technological change for climate change policy discussed in greater detail below.

The preceding discussion highlights several factors that affect the viability of state climate change regulation, including a state Environmental Competition Statute, to mitigate GHG emissions. First, the small scale of state programs severely limits their ability to meaningfully affect global GHG emission levels. Second, leakage may overwhelm regulatory programs, particularly larger ones that involve mobile industries that manufacture undifferentiated goods – the textbook case being reshuffling of electric power generation between existing sources. Third, leakage will be minimal and certainly will not nullify state-level GHG reductions if the regulated industries cannot easily relocate or if they produce differentiated goods. Taken together, these factors indicate that the net effect of leakage will often be neutral or modest, implying that the impacts on GHG emissions will not undermine other grounds for state action. Further, if a federal law, such as a cap-and-trade regime, were promulgated that covered the relevant sources, it would further mitigate, if not eliminate, emissions leakage between states.

Technological Change as a Primary Objective of Climate-Change Policies

Innovation will be an essential element of efforts to mitigate climate change, as well as critical to controlling the costs of climate change policies. Yet technological change continues to be overshadowed by policy makers' and public interest in direct commitments to reducing GHG emissions. This bias is clearly evident in debates over the merits of state programs for mitigating climate change, which have focused on the capacities of cap-and-trade programs and environmental taxes to reduce GHG emissions.[16] In this section, I aim to correct this oversight and to examine the potential for state policies to facilitate the technological transformations that will be essential to mitigating climate change.

Technological change encompasses research and development that produces new technologies and adoption (or diffusion) of existing technologies, which can itself produce innovation through "learning by doing" as experience is gained with the use and production of a technology.[17] This section will highlight the significant potential of technology adoption and learning by doing, both of which an Environmental Competition Statute might spur.

One of the most prominent studies on climate change mitigation identifies a collection of fifteen technology-specific stabilization wedges that can be used to stabilize carbon emissions over the next fifty years. While acknowledging the importance of new, groundbreaking technologies, the authors show that existing technologies are more than adequate for meeting standard stabilization goals through about 2050.[18] These findings provide a strong basis for the importance of promoting technology adoption to climate change mitigation. They also demonstrate that calls for policies to advance technology adoption are not based on mere speculation; many effective technologies already exist.

Learning by doing is derivative of technology adoption. It originates from the observation that the attributes and production of existing technologies improve and evolve as experience is gained with them (the rapid evolution of computer programs following their commercialization is exemplary of this process). Technology assessments, such as the stabilization-wedges analysis, incorporate assumptions about innovation rates from learning by doing. Significant knowledge gaps remain, however, concerning actual rates of innovation from learning by doing. Data for the energy sector, for example, suggest that learning rates are significant and that the corresponding cost savings when integrated over time are in the billions of dollars.[19] As even this example shows, the existing studies are far from comprehensive, though, and often fragmentary.

Predictably, not everyone is convinced by the available data and theories, including some prominent commentators who have challenged the significance of innovation from learning by doing for climate change mitigation.[20] I expect that innovation from learning by doing will be significant for many technologies relevant to mitigating climate change (e.g., solar power, energy-saving technologies), but given the importance that simply adopting existing technologies has for achieving GHG emission reductions, my argument is not dependent on it.

Sources of Technology Market Failure

Economists have long recognized that technological innovation is subject to market failure stemming from inventors' inability to appropriate the full social value of their work. This lost value leads to underinvestment in research and development. Technology adoption and the innovation from learning by doing that it promotes are impeded for similar reasons. Early adopters of new technologies, for example, absorb the costs of working through the kinks in early versions of a technology. This learning process produces valuable knowledge and refinements, which firms adopting the technology later benefit from without having to incur any of the costs. These knowledge spillovers, which early adopters cannot fully internalize, cause investment in learning by doing to be socially suboptimal.

Government regulation has the potential to offset technology market failures because "the rate and direction of innovation are likely to respond to changes in relative prices."[21] Under this theory, governments can induce technological change either directly, through subsidies, or indirectly, through increasing potential market payoffs. Accordingly, government policies that increase the costs of polluting activities or improve the economics of innovative work can potentially stimulate innovation that lowers pollution abatement costs and, in doing so, enhance collective capacities to reduce GHG emissions.

An Environmental Competition Statute has attributes of both an indirect enhancement in market payoffs and a traditional environmental tax on GHG emissions. As we will see more fully below, it is the Environmental Competition Statute's effective increase in marginal profits that is most important to stimulating innovation and offsetting technology market failures, although it is also true that the combination of such complementary measures creates important synergies that enhance the effectiveness of the technology and regulatory policies built into an Environmental Competition Statute.

Policies will differ, however, according to whether they are designed to promote adoption of existing technologies or development of new ones. For companies considering whether to invest in developing new technologies, the key factors will be research and development costs, expected revenues, projected market share, and any likely royalties. The latter three factors are all strongly tied to the size of the market, which implies that a national innovation policy will be favored over a limited state-based one. By contrast, adopters of existing technologies will typically focus on factors such as capital and operating costs, product characteristics, and the environmental benefits of a product. Because none of these factors is sensitive to market size, inducing technology adoption can occur at any level of government.

Differences between developing a new technology and adopting an existing technology prove critical to preserving a role for state-level policies. Because companies developing new technologies will often be different from those adopting them, the market factors that affect technology diffusion, as well as learning by doing, will be distinct from those relevant to promoting research and development. Further, although market size will limit the capacity of environmental policies to spur investments in research and development – the larger a regulated market, the larger the effective incentive – it is irrelevant to states' capacities to induce technology adoption.

State-level policies also have certain advantages over their federal counterparts. Although a federal standard can reach a much greater number of potential technology adopters, multiple state-level measures can mitigate problems with tunnel vision, pork-barrel politics, and picking the wrong technology that can compromise technology programs. For instance, the rise in federal support for biofuels, particularly ethanol produced from corn, is a glaring example of interest-group politics overtaking sound policy.[22]

State programs can generate a diversity of approaches by virtue of their multiplicity and differing mixes of socioeconomic, environmental, and political factors. For example, in the field of renewable energy, some states require that solar power constitute a specific share of an electricity provider's portfolio, while others emphasize wind or geothermal resources.[23] Similarly, states such as West Virginia and Ohio, both of which have large supplies of coal, are supporting innovation directed at clean coal technology,[24] whereas Texas, with its abundant wind resources, has focused on generating power from wind turbines.[25] Other states, such as New Jersey, have been driven by a mix of the potential threats (e.g., sea-level rise) and economic considerations (e.g., impacts on its large chemical production industry) and adopted a more integrated strategy that does not favor a single technology or approach.[26] The variation in state-level programs thus reflects the diversity of conditions present in the states.

The case for state action on climate is bolstered further by the diseconomies of scale endemic to technological change, whether innovation of new technologies or adoption of existing ones. In particular, whereas meaningful reductions in GHG emissions require coordinated large-scale action, technological change often occurs most readily at small geographic scales. Broad consensus exists that innovation is enhanced in geographic clusters (i.e., the Silicon Valley phenomenon) because spatial concentrations allow inventors and adopters to access knowledge externalities that reduce the costs of research, development, and commercialization.[27]

States clearly have a role to play in promoting technological change. To the extent that market size matters, state programs will be inferior to federal regulation. However, although state-level regulation may provide weaker overall incentives, its compensating virtue is the diversity of approaches and experimentation that are a hallmark of state policies. Moreover, where innovation is subject to substantial uncertainties, diversity is often more important than the coordination and large scale found in federal programs.[28] These competing factors reveal important trade-offs between federal and state programs, particularly as they apply to research and development. By contrast, inducing technology adoption, which is insensitive to the size of the market being regulated, is less constrained by these trade-offs. Finally, states are arguably in a better position to establish geographically concentrated centers of innovation that can boost development of new technologies. A state Environmental Competitiveness Statute, particularly through its enhanced market payoffs, can thus play an important role in promoting technological adoption critical to addressing climate change.

Research and Development: The Temporal Schism in Technology Forcing

Limited empirical support exists for the effectiveness of environmental regulation, at any level of government, inducing research and development. Two recent studies have identified modest correlations between pollution abatement costs and either: (1) patenting rates in related technological areas or (2) levels of research and development.[29] More extensive studies exist for research and development related to energy-efficient technologies. Researchers have, for example, observed "significant amounts of innovation" in response to both increases in energy prices and changes in energy-efficiency standards.[30]

None of these studies provides compelling evidence that environmental regulation offers a powerful means of stimulating research and development. This is not to suggest that regulations can never be effective in this respect. To the contrary, the 1970 Clean Air Act successfully spurred significant research and development.[31] But even this example is subject to qualification, as the extended grace period the statute provided and fortuitous timing of the new technology were critical to the success of the regulations.

The equivocal empirical data may reflect deeper problems. The degree to which environmental regulations can promote technological innovation is hampered by the political economy of the regulatory process itself. This critique turns on the realization that incentives to conduct research and development, because of the long lag time between conducting research and sale of a product, are tied to the stringency of regulations in place when a commercially viable product is produced. Consequently, under this view, it is the credible threat of stringent regulation in the future – not primarily policies currently in place – that is most relevant to spurring investments in research and development.[32]

The extended lag between investment decisions and the relevant regulatory policies creates uncertainties that erode regulatory incentives. These uncertainties stem from the disparities that exist between current and future government policies. In fact, the optimal policy today – permit levels that reflect the costs of adoption *and* invention – will not be the politically expedient or economically optimal policy in the future when regulatory standards matter.[33] The current debate over drug pricing is perhaps the best example of this dynamic. Politicians and consumers want to reduce drug prices to just above the marginal cost of producing them whereas drug companies object that this ignores the high costs of research and development. There are many nuances to this debate, on both sides, but the central tension is the same regardless of the underlying technology. Moreover, the higher the political or moral stakes, typically the greater the pressure for producers to lower their prices. These uncertainties are compounded by those associated with the projected magnitude and timing of climate change and the corresponding societal impacts.

The questionable credibility of government commitments to future levels of regulation diminishes the capacity of environmental regulations to induce companies to invest in long-term research and development. This temporal paradox has led some observers to conclude that environmental regulation has little or no capacity to induce innovation relevant to mitigating climate change. For them, the benefits of environmental regulations rest solely on their capacity to "minimize the costs of technology adoption for *existing* technologies" or to stimulate short-term innovation with "immediate implications for cost savings."[34] Insofar as an Environmental Competition Statute relies on policies that provide incentives that accrue in the future, which is true of both its regulatory and subsidy provisions, it is susceptible to this critique.

These arguments expose the pitfalls of ignoring the political (and economic) realities of environmental regulation at any level of government. One must be careful, however, not to take this argument too far. For one, this critique runs contrary to those studies and examples suggesting that environmental regulation can influence research and development. More fundamentally, the critique rests on a narrow conception of economic rationality that does not necessarily match common understanding of political decision-making processes. Little doubt nevertheless exists that the politics of regulation and uncertainties in climate change science diminish the incentives that environmental regulation – whether state or federal – can provide for investment in extended research and development. One admittedly modest way to mitigate these uncertainties is to include phase-in provisions that match the expected time (often decades) for new technologies to be developed. Alternatively, in the case of an Environmental Competition Statute, it could be structured to pull some of the revenue it generates for upfront direct subsidies similar to traditional government research and development programs.

Adoption of Existing Technologies: A More Promising Role for the States

Significant empirical support exists for the benefits of using environmental regulations to induce technology adoption. Examples include studies of prominent regulations under the Clean Air Act and the Clean Water Act.[35] Positive correlations have also been found between energy prices and adoption rates of energy-efficient products.[36] The most compelling evidence is associated with Corporate Average Fuel Economy (CAFE) standards, which have been found to be substantially more effective than increases in fuel prices.[37] Consistent with these findings, researchers have found that "technology adoption decisions are more sensitive to up-front costs than to longer-term operating expenses."[38]

One potential reason for these robust results is the absence of a temporal schism between the timing of regulation and investments in technology adoption. Both technology adoption and learning by doing are responsive to incentives created by current environmental policies. This dramatically reduces the uncertainties that undermine regulatory incentives for innovation, although they are by no means eliminated because the policies themselves are, of course, subject to change by state (or federal) governments. On the other hand, competing companies must comply with a regulation, meaning that no one is at a competitive disadvantage by adopting an underused technology – even if a state later chooses to relax its regulatory standards – and free-rider problems are minimized because learning by doing occurs in parallel.

Paradoxically, the successes in using environmental regulations to promote technology adoption may also owe something to the many barriers that impede it. Many studies have shown that adoption of new technologies is slow even when they are clearly superior to existing technologies with respect to cost and performance. Economists, for example, have calculated implicit discount rates for consumers purchasing energy-efficient technologies that far exceed market interest rates (25 percent and often much higher).[39] As these findings suggest, multiple potential barriers exist including transaction costs, bounded rationality, information deficits, technological and financial risks, and investor-user splits.[40] Unfortunately, the factors that impede adoption of specific technologies are highly variable across different classes of technologies and no unifying theory currently exists to guide policy.[41] Policy makers must therefore be attentive to the specific barriers that may be present (e.g., onerous upfront capital costs) and structure policies accordingly.

Technology adoption and learning by doing has a significant local component that shrinks the scale at which success is measured. Both, as argued previously, are enhanced by knowledge externalities captured only in geographically localized communities. Further, the trajectory of technology adoption typically follows an S-curve – rising gradually at first, then steeply increasing and finally leveling off – such that success is marked by achieving levels of adoption that trigger a shift to rapid diffusion (the steep part of the adoption curve).[42] State-level programs, such as renewable portfolio standards for electric power generation,[43] thus have distinctive advantages over federal programs insofar as they can be more precisely targeted to meet specific barriers to technology adoption and can sustain a diversity of approaches. State-level Environmental Competition Statutes should therefore focus on promoting technology adoption and learning by doing over innovation.

A Hybrid Approach to Climate Change Policy

One of the great virtues of federal climate change policies, particularly when linked to an international regime, is that they can overcome the eroding, or even perverse, effects of emissions leakage on GHG reductions. I have shown already how, in the absence of a federal program or where gaps in a federal regime may exist, states can adopt technology-oriented policies that circumvent the limited scale of state programs and their vulnerability to emissions leakage. This strategy sidesteps the challenges of achieving meaningful reductions in GHG emissions by focusing on technological change. In this section, I address the circumstances in which states operate under the backdrop of significant federal legislation, such as a multisector cap-and-trade program.

An important potential limitation of this approach is the irreducible uncertainties that may foreclose standard optimization strategies. The efficiency gains, as opposed to benefits alone, of induced technological change have proved difficult to demonstrate. William Nordhaus and other researchers have estimated that "the impact of induced innovation is modest."[44] By contrast, other studies have predicted cost savings from induced innovation of 50 percent,[45] and where cost functions are convex (i.e., the rate of increase rises with required abatement level) and the abatement standard is stringent, another study has found the benefits of induced innovation to be dramatic.[46] These inconsistencies suggest that the best we may be able to achieve is increases in relative efficiency, often referred to as satisficing, as opposed to optimization.

The available studies provide solid grounds for identifying state policies that meet this satisficing principle. A recent set of studies evaluates strategies that combine market-based environmental regulations with technology-push policies. The estimated benefits over either type of approach on its own are impressive. As one might expect, given the long lag times and high costs of research and development, a dual approach is far superior (about a factor of ten less costly) to a straight technology policy.[47] Although not quite as dramatic, dual approaches that combine a carbon tax with direct innovation subsidies can reduce costs by more than one-third over a carbon tax on its own.[48] These studies provide strong support for policies that utilize complementary measures, such as a carbon tax combined with subsidies for innovation and learning by doing. A hybrid approach can also enhance the efficiency of other measures, such as renewable portfolio standards, by implementing them in parallel with, for example, market-based regulations.[49]

By allowing environmental and technology objectives to be decoupled, a hybrid approach circumvents the tensions created by these sometimes-competing goals. In particular, near-term GHG reductions can be pursued efficiently, if focused on technology adoption, because they are not burdened by the perceived need to implement stringent standards to spur investment in costly research and development. At the same time, by recognizing the limitations of using regulations to stimulate research and development, one can leverage technology policies to subsidize research and development. This strategy has the virtue of ensuring that new technologies will be available in the future while limiting abatement costs and encouraging continual improvements in technologies and their production processes over time. A federal cap-and-trade program, for example, aimed at reducing GHG emissions could be combined with state Environmental Competition Statutes aimed at promoting technology adoption and learning by doing.

The efficiencies gained by combining environmental regulatory measures and technology policies is borne out by a recent comparative study of environmental taxes in Europe targeted at reducing carbon dioxide emissions. The author, Monica Prasad, found that environmental taxes in several Scandinavian countries dating back to the early 1990s have had little discernable effect, except in the case of Denmark.[50] Prasad identified several reasons for Denmark's success including, most importantly for our purposes, that "Danish policy makers made huge investments in renewable energy and subsidized environmental innovation."[51] By ensuring that substitute technologies were readily available, the Danish government overcame the risk aversion of firms, countervailing sunk costs, and network effects that often limit technology adoption. Denmark encouraged technology adoption through direct subsidies, differentially high taxes on coal, and tax benefits to industries that voluntarily agreed to reduce emissions. Prasad concluded that this mix of environmental taxes and technology policies was essential to Denmark's success and ultimately distinguished its system from the failed policies of other European countries.

The existing econometric analyses and Prasad study reveal the benefits of combining traditional environmental regulations with technology policies to address climate change. They also highlight the importance of remaining attentive to the two distinct but related objectives of climate policy – reducing GHG emissions and promoting the technological changes that will be essential to meeting long-term GHG emissions targets.

The preceding sections have explored the strengths and weakness of different regulatory and technology policies in an effort to identify the policies most amenable to state action when a federal regime is in place. This exploration leads to three central conclusions. First, state climate change regulations should not center on reducing GHG emissions. Second, the primary objective of state climate change policies, including any state-level Environmental Competition Statutes, should be to induce technological change, principally through technology adoption and innovation through learning by doing. Third, complementary regulatory and technology policies should be adopted by states, with particular emphasis on market-based regulations, technology portfolio standards, and innovation subsidies, while the federal government retains primary responsibility for direct regulation of GHG emissions.

Recommendations: Environmental Federalism in an Era of Climate Change

A central premise of this chapter is that federal and state climate change policies should be complementary. Federal action is essential to achieving the dramatic GHG emissions reductions that scientists predict will be necessary, and it will play a critical role in promoting technological change as well. The primary role of the states, as I have argued previously, should be fostering technological change, particularly technology adoption and innovation through learning by doing. Federal laws should therefore enhance the ability of states to contribute to climate change mitigation in these ways. With this objective in mind, I propose several recommendations for federal climate change legislation.

State and Regional Cap-and-Trade Programs Should Be Retained but Should Not Be the Focus of State Climate Efforts

Much ink has been spilled analyzing the ways in which a federal cap-and-trade regime could be harmonized with state and regional programs, such as RGGI and California's A.B. 32. As I have shown, once a federal program enacts a legitimate cap-and-trade program, investment of substantial resources in parallel state or regional cap-and-trade programs is unwarranted if the objective is reducing GHG emissions, as opposed to promoting new policy development. These programs, particularly if focused on the short-term, generate little in the way of technology development or adoption, and they achieve only nominal reductions in GHG emissions relative to even a modest federal program.

Federal legislation nonetheless should not entirely preempt state and regional cap-and-trade programs. State and regional GHG emissions reduction programs have the potential to augment emissions reductions over and above those of any federal program that is ultimately adopted. Although the amount will be small, federal programs are unlikely to be optimally stringent and any additional reductions should not be needlessly precluded.

The policy development and institutional capacity building generated by state GHG regulations offer further grounds for preservation. The RGGI and A.B. 32 have already made important contributions to policy development by requiring participants to work through the details of regulatory regime and functioning as a proverbial laboratory of democracy. These contributions will not end on the enactment of federal legislation regime if for no other reason than that federal and state programs differ in many important ways. For instance, although each of the leading climate change bills in Congress allocate most of the GHG allowances without charging a fee,[52] RGGI gives each member state broad discretion in determining whether it will auction off permits.[53] Currently, five states (Maine, Massachusetts, New York, Rhode Island, and Vermont) have pledged to auction off 100 percent of their state's emissions allowances.[54] Preservation of state programs is particularly important given the complexity of these programs, as it will enable much greater experimentation than a unitary federal system.

Federal and State Climate Change Programs Should Adopt Complementary GHG Emissions-Reducing and Technology-Forcing Measures

The importance of technological development to climate change policy and the distinct advantages of pursuing multiple policies to induce technological change provide a strong basis for a continuing state role in climate policy making, even after the enactment of federal climate legislation. By adopting a complementary approach, states and the federal government leverage synergistic benefits that are greater than either can achieve alone. Under this approach, emissions reductions should be left primarily to federal policy (i.e., cap-and-trade regimes), while states focus on technology change in parallel with the federal government. An Environmental Competition Statute is well suited to this framework, as it has the ability to provide significant incentives for companies to promote technology adoption and learning by doing.

A complementary approach is also consistent with standard principles of federalism. It reflects the understanding that some policies are more effectively pursued by the federal government, while others are better addressed by the states. Beyond their focus on reducing GHG emissions, federal policies can make valuable contributions by establishing minimum technology portfolio standards, regulating sectors that transcend state boundaries (e.g., transportation), and offering subsidies for research and development and technology adoption. The states are well suited to playing a complementary role through measures, particularly Environmental Competition Statutes, that promote adoption of existing technologies or learning by doing or that provide direct subsidies for technological change generally.

NOTES

1. Pew Ctr. on Global Climate Change, The American Clean Energy and Security Act (Waxman-Markey Bill) (May 2009), *available at* http://www.pewclimate.org/acesa.
2. *See generally* Pew Ctr. for Global Climate Change, *What's Being Done in the States*, *available at* http://www.pewclimate.org/what_s_being_done/in_the_states; Barry G.

Rabe, Statehouse and Greenhouse: The Emerging Politics of American Climate Change Policy (2004).

3. Adam Jaffe et al., *A Tale of Two Market Failures: Technology and Environmental Policy*, 54 Ecological Econ. 164, 168–69 (2005).
4. James Bushnell et al., *Local Solutions to Global Problems: Climate Change Policies and Regulatory Jurisdiction*, 2 Rev. Envtl. Econ. & Pol'y 175, 192 (2008), *available at* http://reep.oxfordjournals.org/cgi/reprint/ren007v1; Jonathan B. Wiener, *Think Globally, Act Globally: The Limits of Local Climate Policies*, 155 U. Pa. L. Rev. 1961, 1972–73 (2007); Lawrence H. Goulder, *California's Bold New Climate Policy*, Economist's Voice 1, 1 (Sept. 2007).
5. Robert W. Hahn, The Economics and Politics of Climate Change 44–54 (1998); Kirsten H. Engel & Scott R. Saleska, *Subglobal Regulation of the Global Commons: The Case of Climate Change*, 32 Ecology L.Q. 183, 223–26 (2005).
6. Daniel C. Esty, *Toward Optimal Environmental Governance*, 74 N.Y.U. L. Rev. 1495, 1555 (1999); Robert N. Stavins, *Policy Instruments for Climate Change: How Can National Governments Address a Global Problem?* 1997 U. Chi. Legal F. 293, 323; Wiener, *supra* note 4, at 1972–73 (2007).
7. To the extent it has been considered, the focus has been on determining the appropriate policy instruments. *See, e.g.*, David M. Driesen, *Design, Trading, and Innovation*, in Moving to Markets in Environmental Protection: Lessons after 20 Years of Experience 436, 436 (Jody Freeman & Charles D. Kolstad eds., 2007).
8. Adam B. Jaffe et al., *Technological Change and the Environment*, *in* 1 Handbook of Environmental Economics 461, 476 (Karl-Goran Maler & Jeffrey R. Vincent eds., 2003).
9. *See, e.g.*, Martin I. Hoffert et al., *Advanced Technology Paths to Global Climate Stability: Energy for a Greenhouse Planet*, 298 Science 981, 981 (2002); S. Pacala & R. Socolow, *Stabilization Wedges: Solving the Climate Problem for the Next 50 Years with Current Technologies*, 305 Science 968, 968 (2004).
10. Reg'l Greenhouse Gas Initiative, *About RGGI*, *available at* http://www.rggi.org/about.htm. The committed states include Connecticut, Delaware, Maine, Maryland, New Hampshire, New Jersey, New York, and Vermont. The District of Columbia and three states, Massachusetts, Rhode Island, and Pennsylvania, are seriously considering joining RGGI. *Id.*
11. *See* EPA, State CO2 Emissions from Fossil Fuel Combustion, 1990–2005, *available at* http://www.epa.gov/climatechange/emissions/downloads/CO2FFC_2005.pdf.
12. Corrado Di Maria & Edwin van der Werf, *Carbon Leakage Revisited: Unilateral Climate Policy with Directed Technological Change* 2–3 (Fondazione Eni Enrico Mattei, Working Paper No. 94.2006), *available at* http://ssrn.com/abstract=912461.
13. Mustafa H. Babiker, *Climate Change Policy, Market Structure, and Carbon Leakage*, 65 J. Int'l Econ. 421, 441–43 (2004).

14. Di Maria & van der Werf, *supra* note 12, at 16; Rolf Golombek & Michael Hoel, *Unilateral Emission Reductions and Cross-Country Technology Spillovers*, 4 J. ADVANCES ECON. ANAL. POL'Y 1, 18–19 (2004).
15. Rabe, *supra* note 2, at 169. Diffusion of climate change policies and knowledge is also occurring across national borders. PEW, *supra* note 1, at 6.
16. Lawrence H. Goulder & Stephen H. Schneider, *Induced Technological Change and the Attractiveness of CO_2 Abatement Policies*, 21 RESOURCE & ENERGY ECON. 211, 212–13 (1999).
17. Richard G. Newell et al., *The Effects of Economic and Policy Incentives on Carbon Mitigation Technologies*, 28 ENERGY ECON. 563, 564–66 (2006).
18. Pacala & Socolow, *supra* note 9, at 968–69.
19. Alan McDonald & Leo Schrattenholzer, *Learning Rates for Energy Technologies*, 29 ENERGY POL. 255, 256 (2001).
20. W. David Montgomery & Anne E. Smith, *Price, Quantity, and Technological Strategies for Climate Change Policy*, in HUMAN-INDUCED CLIMATE CHANGE: AN INTERDISCIPLINARY ASSESSMENT 332 (M. Schlesinger et al. eds., 2005).
21. Jaffe et al., *supra* note 8, at 469–70.
22. Alexei Barrionuevo, *Boom in Ethanol Reshapes Economy of Heartland*, N.Y. TIMES, June 25, 2006, at A1.
23. *See generally* Database of State Incentives for Renewables and Energy Efficiency, DSIRE Home, *available at* http://www.dsireusa.org.
24. PEW, *supra* note 1, at 9.
25. Rabe, *supra* note 2, at 60–61.
26. *Id.* at 110–11.
27. Maryann Feldman & Roger Martin, *Constructing Jurisdictional Advantage*, 34 RES. POL'Y 1235, 1236–37 (2005); David B. Audretsch & Maryann P. Feldman, *Knowledge Spillovers and the Geography of Innovation*, *in* 4 THE HANDBOOK OF URBAN AND REGION ECONOMICS 2713, 2718 (J. Vernon Henderson & Jacques-Francois Thisse eds., 2004).
28. Robert P. Merges & Richard R. Nelson, *On the Complex Economics of Patent Scope*, 90 COLUM. L. REV. 839, 873 (1990).
29. Adam B. Jaffe & Karen Palmer, *Environmental Regulation and Innovation: A Panel Data Study*, 79 REV. ECON. & STATS. 610, 611–12 (1997); Jean Olson Lanjouw & Ashoka Mody, *Innovation and the International Diffusion of Environmentally Responsive Technology*, 25 RES. POL'Y 549, 550–51 (1996).
30. Jaffe et al., *supra* note 8, at 475.
31. *See* NAT'L RESEARCH COUNCIL, COMM. ON STATE PRACTICES IN SETTING MOBILE SOURCE EMISSIONS STANDARDS, STATE AND FEDERAL STANDARDS FOR MOBILE-SOURCE EMISSIONS 116 (2006). *See also* Thomas O. McGarity, *Radical Technology-Forcing in Environmental Regulation*, 27 LOY. L.A. L. REV. 943 (1994).
32. Montgomery & Smith, *supra* note 20, at 7–8; Jean-Jacques Laffont & Jean Tirole, *Pollution Permits and Environmental Innovation*, 62 J. PUB. ECON. 127, 128–29 (1996).

33. Montgomery & Smith, *supra* note 20, at 27; Carolyn Fischer & Richard G. Newell, *Environmental and Technology Policies for Climate Mitigation*, 55 J. ENVTL. ECON. MGMT. 142, 144 (2008).
34. Montgomery & Smith, *supra* note 20, at 6–7.
35. Jaffe et al., *supra* note 8, at 502; Suzi Kerr & Richard G. Newell, *Policy-Induced Technology Adoption: Evidence from the U.S. Lead Phasedown*, 51 J. IND. ECON. 317, 340–41 (2003).
36. Richard G. Newell et al., *The Induced Innovation Hypothesis and Energy-Saving Technological Change*, 114 Q.J. ECON. 941, 967–70 (1999); Newell et al., *supra* note 17, at 567.
37. Patricia K. Goldberg, *The Effects of the Corporate Average Fuel Economy Standards in the U.S.*, 46 J. IND. ECON. 1, 2–3 (1998); David L. Greene, *CAFE or Price? An Analysis of the Effects of Federal Fuel Economy Regulations and Gasoline Price on New Car MPG, 1978–89*, 11 ENERGY J. 37, 55–57 (1990).
38. Jaffe et al., *supra* note 8, at 496; *see also* Newell et al., *supra* note 17, at 571.
39. Richard B. Howarth & Bo Anderson, *Market Barriers to Energy Efficiency*, 11 ENERGY ECON. 262, 262–63 (1993); Newell et al., *supra* note 17, at 570–71.
40. Joachim Schleich & Edelgard Gruber, *Beyond Case Studies: Barriers to Energy Efficiency in Commerce and the Services Sector*, 30 ENERGY ECON. 449, 453–55 (2008). Transaction costs include obtaining relevant information, negotiating contracts, and so on; bounded rationality limits agents' ability to optimize because of lack of time, attention, or capacity; limited access to capital may cause adoption to be unprofitable because capital is so expensive; the investor-user dilemma arises when, for example, neither a renter nor a landlord has the incentive to invest in energy-saving technologies because he or she cannot recoup the costs; and technological or financial risk creates significant uncertainty about potential returns on an investment. *Id.*
41. Timothy F. Malloy & Peter Sinsheimer, *Innovation, Regulation and the Selection Environment*, 57 RUTGERS L. REV. 183, 190–91, 233 (2004); P.A. Geroski, *Models of Technology Diffusion*, 29 RES. POL'Y 603, 621 (2000).
42. Geroski, *supra* note 41, at 604.
43. *See, e.g.*, JOHN A. ALIC ET AL., PEW CENTER FOR GLOBAL CLIMATE CHANGE, U.S. TECHNOLOGY AND INNOVATION POLICIES: LESSONS FOR CLIMATE CHANGE 13 (2003), *available at* http://www.pewclimate.org/global-warming-in-depth/all_reports/technology_policy.
44. Jaffe et al., *supra* note 8, at 484; *see also* Ian W.H. Parry et al., *How Large Are the Welfare Gains from Technological Innovation Induced by Environmental Policies?* 23 J. REG. ECON. 237, 239 (2003).
45. Hadi Dowlatabadi, *Sensitivity of Climate Change Mitigation Estimates to Assumptions about Technical Change*, 20 ENERGY ECON. 473, 473–74 (1998).
46. LAWRENCE H. GOULDER, PEW CTR. ON GLOBAL CLIMATE CHANGE, INDUCED TECHNOLOGICAL CHANGE AND CLIMATE POLICY, 14–15 (2004), *available at* http://www.pewclimate.org/global-warming-in-depth/all_reports/itc. Econometric estimates of

the benefits from learning by doing are similar in magnitude, with one study finding that learning by doing reduces the cost of meeting a 550 parts per million abatement target by 42 percent under a high-cost scenario and by 72 percent under a low-cost scenario. *Id.*

47. Stephen H. Schneider & Lawrence H. Goulder, *Achieving Low-Cost Emissions Targets*, 389 NATURE 13, 14 (1997).
48. Fischer & Newell, *supra* note 33, at 144.
49. *Id.* at 143.
50. Monica Prasad, *Taxation as a Regulatory Tool: Lessons from Environmental Taxes in Europe* 4–5, *available at* http://www.sociology.northwestern.edu/faculty/prasad/Taxation_3_25_08; Monica Prasad, *On Carbon, Tax and Don't Spend*, N.Y. TIMES, Mar. 25, 2008, at A27.
51. *Id.* at A27.
52. *See* Pew Ctr. for Global Climate Change, *Table 1: Selected Allocation Approaches Proposed in the 110th Congress*, *available at* http://www.pewclimate.org/docUploads/images/Table%201%20image.bmp.
53. Debra Kahn, *Northeast States Prep for Inaugural Carbon Auction*, GREENWIRE, Jan. 22, 2008, *available at* http://www.earthportal.org/news/?p=823.
54. *Id.*

10 The iUtility

Joseph P. Tomain

COAL-FIRED POWER PLANTS IN THE UNITED STATES ARE RESponsible for 10 percent of the world's carbon dioxide emissions.[1] Consequently, any strategy aimed at reducing carbon emissions must address the electric industry. The electric industry, in turn, must reform itself, and the single most significant reform involves a change in an electric utility's mission. Historically, electric utilities were driven by one concern and one concern only – sell electricity. The mission was simple and understandable. Electricity consumption supported the economy: QED, the more electricity that was sold and consumed, the stronger the economy was. The traditional mission served the country very well for most of the twentieth century. Our economy was healthy and expanding; an electricity infrastructure was constructed and reliable; and electricity was abundant and cheap.

The electricity industry, except in its very nascent stage, never operated in a free, competitive market. Rather, it has been supported by a regulatory compact between government and industry. The regulatory compact, and the parties to it, must change to meet the demands of the new environmentalism. Government regulation must change to promote the public good of a cleaner environment. Utilities must change to sell energy services and products – not just sell electricity.

An Environmental Competition Statute (ECS) for power producers can assist forward-thinking utilities interested in investing and developing new technologies along three paths. The intelligent utility will (1) reduce carbon

This chapter is an expanded version of Joseph P. Tomain, *Building the iUtility*, 146 Pub. Util. Fort. 28 (August 2008).

emissions, (2) increase energy efficiency, and (3) produce clean electricity. The ESC's structure of transfer payments from traditional dirty utilities to innovative utilities, along these three lines, can support these efforts to respond to the challenges of climate change.

This chapter focuses on how electric utilities and the regulations governing them must be restructured. The mission of today's smart utility must be to produce energy that is environmentally sensitive. The simple-sounding shift in mission from focusing on power production and the economy to focusing on power production, the economy, and the environment cuts against more than one hundred years of utility regulation. A transition from a traditionally structured utility to a modern, smart utility will change the nature of an electric utility's business, from its product mix to its pricing policies.

The iPhone serves as model and metaphor for the new electric utility – the iUtility. The iUtility resembles a smoke-belching power plant no more than the iPhone resembles the heavy black rotary-dial telephone of the past. Rather, the iUtility, like the iPhone, is technologically sophisticated; offers various services and products at various prices; and can be customized, decentralized, and personalized as distinguished from the large-scale, centralized power plants of the past. This chapter describes the policy assumptions and the regulatory history behind the traditional regulatory compact and demonstrates why the future requires a new compact based on new assumptions.

This chapter then explains the assumptions behind a new energy policy and also explains how the iUtility responds to those assumptions and can contribute to the next generation of environmental law by reconfiguring the regulatory compact that has long governed the industry. It then discusses how the incentives created by the ECS would coincide with and help achieve the policy goals of the new regulatory compact.

U.S. Energy Policy and Profile

The United States has no coordinated or comprehensive energy policy, despite the fact that the Department of Energy regularly reports on existing energy policies and practices and occasionally a national energy plan is drafted, the most recent of which appeared in May 2001.[2] Although it is a misnomer to use the phrase "energy policy," U.S. policy can be described in broad contours.

U.S. energy policy is based on certain assumptions. We assume that free markets, private ownership, and private investments are preferable to central energy planning for social ordering. We also assume that energy industries should be treated separately rather than be coordinated. And we assume that there is a direct and positive correlation between energy production and consumption, and between energy production and economic growth. In short, these assumptions lead to the conclusion that bigger is better (and cheaper). This conclusion is correct – to a point.[3] Large energy producers can achieve economies of scale and produce energy more efficiently, which is to say more cheaply, particularly when they avoid paying the full social costs of pollution.

As a result of these assumptions, our energy industries have developed certain characteristics. The oil, natural gas, nuclear power, coal, and electricity industries all share the characteristics of being large scale, capital intensive, and highly centralized. In other words, our energy industries are national and global in scope rather than local and decentralized. These industry characteristics would be beneficial and benign were it not for the large-scale fouling of the commons.

The U.S. energy profile can be roughly divided between oil and electricity. This division of our energy economy into these two sectors is significant. More than two-thirds of our oil consumption is devoted to transportation, with slightly more than 1 percent used to generate electricity.[4] Electricity is used in the industrial, residential, and commercial sectors largely for heating, cooling, and lighting, with roughly 0.2 percent used in the transportation sector.[5] Half of all electricity is generated by coal, followed by roughly 20 percent generated by natural gas and 20 percent by nuclear power. The remaining electricity is generated by hydropower and other resources such as wind, solar, geothermal, and biomass.[6] Fossil fuels (coal and natural gas), then, account for most of the electricity generated in the country, and it is the electricity industry that contributes about 40 percent of all U.S. carbon dioxide emissions.[7] More significantly, coal is abundant and cheap at between $1 and $2 per million Btu, as compared with between $6 and $12 per million Btu for natural gas and oil.[8] Consequently, the electricity industry is a major source of the pollutants that contribute to climate change. The ECS can be applied to reduce the impact of carbon dioxide and other greenhouse gases emitted into our atmosphere. Consequently, the statute can play a significant role in assisting utilities to change their method of doing business as usual and become active participants in the green economy.

Industry Overview

The electricity industry, a roughly $300 billion industry,[9] has unique characteristics. There are approximately 4,000 electric power suppliers, 203 of which are known as investor-owned utilities (IOUs). The IOUs are privately owned and generate 72 percent[10] of all electricity, whereas public entities, both federal and local, produce the remainder.[11] Further, we have developed certain expectations about electricity. Electricity must be continually available. Consumers expect that electricity will be abundant and affordable. Producers expect that large-scale electricity production can achieve significant economies of scale as they have invested billions of dollars in capital plant and infrastructure. And regulators expect that it is their job to facilitate reliable electricity production.

Electricity has its own unique characteristics. Electricity cannot be stored. Just think of how quickly the battery in your laptop or cell phone runs down. Electricity is also a completely fungible or homogeneous good. There is no quality difference between electricity from one source or another. Electricity moves literally at the speed of light, and consequently, no one can identify its point of origin or consumption. In other words, consumers and producers do not identify or contract for specific goods for sale. Instead, producers place electricity onto the electric grid and consumers purchase what they need. Although all electricity performs exactly the same function, electricity service can be more or less reliable; therefore, electricity services vary in price and quality. Further, the second law of thermodynamics dictates that as electricity is produced and transported over distances, entropy will cause a loss of energy, which is known as load loss. There are, thus, efficiency limits to electricity production, transmission, and distribution.

Regulatory History

The electric industry began on September 4, 1882, when Thomas Edison flipped the switch at his Pearl Street Electricity Station, which served fifty-nine customers in the local neighborhood. In 1896, George Westinghouse developed a hydroelectric project at Niagara Falls that served the residents of Buffalo, New York, more than twenty-nine miles away. Electricity firms then grew rapidly in scope, and because of favorable economies of scale, the industry served customers with a declining cost basis for decades. From the turn of the century until the mid-1930s, the industry continued to expand. It also continued to concentrate.

State utility commissions began regulating in the first decades of the twentieth century to combat industry concentration and the exercise of market power. Federal regulation began with the Federal Power Act of 1920, which asserted federal jurisdiction over the nation's hydroelectric resources. In the mid-1930s, the federal government expanded its regulation through holding company regulations and with the regulation of wholesale electricity rates to ensure that they were just, reasonable, and nondiscriminatory. Throughout this period, electric capacity and generation grew at an annual rate of about 12 percent, while electricity prices continued to drop. Also during this period, as the electric industry moved from local to state to regional, and as large hydropower projects came on line, the national electricity grid developed.[12]

From World War II through the 1960s, the electricity industry enjoyed what is known as the golden age.[13] World War II increased demand, privately owned utilities experienced dramatic growth in capacity, and residential and commercial use expanded rapidly. Most notably, utility planners could rely on an annual growth rate of more than 7 percent per year as economies of scale continued to be realized, capacity continued to increase, and prices continued to decline at 1.5 percent annually. The most notable aspect of the great growth of electricity occurred with the passage of the Atomic Energy Act of 1954, which ushered in commercial nuclear power, touted as being too cheap to meter. The promise of nuclear power was that nuclear plants could be built at a large scale and that the primary fuel, uranium, was less expensive than coal. Nuclear power appeared to be the hopeful future of electricity.

The golden age ended in 1965 as electricity production peaked. The marginal cost of electricity exceeded average costs for the first time. In other words, consumers could no longer count on electricity prices declining. In addition, we began to become aware of the environmental harms of electricity production, particularly from coal-fired plants, as industry growth slowed and costs escalated. More trouble was on the way.

The energy crises of the 1970s raised prices across the economy. Power plant construction costs escalated. Environmental laws made the country more pollution conscious. The accident at Three Mile Island in 1979 resulted in the collapse of the nuclear power industry. In short, the characteristics that sustained the industry for the first two-thirds of the twentieth century – industry expansion and declining prices – came to an end. During the last third of the century, the electric industry and its regulators attempted to find calm waters with little success.

Deregulating Electricity

In response to the 1970s energy crises, the National Energy Act was passed in 1978 with a surprising consequence. One portion of the Act, the Public Utility Regulatory Policies Act (PURPA), contained provisions for independent power producers not regulated under traditional utility laws, known as nonutility generators. Later, the Energy Policy Act of 1992 authorized another set of nonutility electricity providers known as exempt wholesale generators. These new entrants could produce cheaper electricity than the incumbent utilities, yet they needed to transmit and distribute their electricity to customers, which meant using the power transmission lines owned by the traditional regulated IOUs – their competitors.

The problem of access to cheaper electricity has plagued the Federal Energy Regulatory Commission (FERC) for decades. Incumbents had sunk billions of dollars in transmission line investments that they did not want to lose to upstarts. Still, FERC responded by attempting to open access and create new electricity markets. Both responses have failed. Once ordered to open access to their transmission lines, incumbents set prices favorable to themselves. In addition, utilities were reluctant to actively participate in FERC's newly designed electricity markets. So FERC decided to require utilities to form independent regional operating companies to buy and sell access, but reliable market design has continued to elude it.

Federal regulations were aimed at the wholesale end of the electricity business, which was relatively competitive and, therefore, subject to deregulation. It logically follows that, if the generation end is more competitive, then the retail end might also benefit from increased competition. The most notable experiment in retail deregulation occurred in California and was a complete failure.[14] The regulatory design for retail competition in California capped retail rates to consumers but required generators to purchase at the market price. The price squeeze is easy to imagine. Rates to retail customers were fixed, but utilities, saddled with an obligation to serve those customers, were required to purchase at whatever market rates were available. At peak times, electricity could be bought only at prices higher than – often times multiples of – normal prices. As a result, California witnessed utility bankruptcies and the trading scandal known as Enron. Enron traders manipulated California electricity markets in manifold ways to create artificial shortages and thus extract supracompetitive prices from the utilities that had to buy the electricity to serve their customers.[15]

The hoped-for deregulation of the electricity industry collapsed dramatically. Instead of achieving competition for the entire industry at both the wholesale and the retail levels, regulators decided to encourage competition for wholesale power producers and to restructure the retail segment of the industry after California, Enron, and the Northeast's blackout of August 2003.[16] Today, restructuring has completely stalled as calls for reregulation are heard. We are left with an electric industry that has enjoyed success in the past and is trying to find itself in a new world. The explanation for the success and the subsequent failure of the industry, as well as a promise of healthier times, requires rethinking traditional electricity regulation.

Traditional Utility Regulation

Traditional utility regulation was based on two complementary ideas – natural monopoly and the regulatory compact. Together, these ideas enabled the industry to grow, to nationalize, and to produce cheap electricity for most of the century. Traditional regulation constructed a utility industry that became increasingly costly to maintain, that hampered new entrants, and that still seems resistant to change. To envision the iUtility requires that we address and rethink both ideas.

Natural Monopoly

Consolidation in the early part of the twentieth century revealed a central fact about the electricity industry – it exercised monopoly market power. Monopolies are economically perverse, as prices are set above competitive levels. Under monopoly conditions, consumers suffer losses that they would not suffer in competitive markets, cheaper producers are prevented from putting their products on line, and society does not maximize the use of its resources. Left unchecked, electric monopolies could, and did, set prices above and reduce supply below the competitive level, thus causing inefficient social losses. But was there any other way to run a utility industry? Perhaps not, as utilities were considered natural monopolies.

The easiest way to understand a natural monopoly is to imagine competing electric firms laying competing sets of transmission and distribution lines throughout the country. It is *the* characteristic of a natural monopoly that one provider can provide the service more cheaply than multiple providers because multiple providers with multiple facilities are simply wasteful.[17] Natural monopoly, then, became the justification for government regulation in the form of the regulatory compact to the great benefit of the industry and its customers for many years.

As a result, the natural monopoly model will need to be either radically reformed or done away with altogether. Currently, the electricity grid contains natural monopoly characteristics. It is not, however, beyond contemplation that, through the extensive use of distributed generation, the grid can be radically transformed and that all utilities can act more as competitors than as monopolists. Competition is the core idea underlying the ECS. The statute can encourage new entrants through technological innovation and promote competition among them, and it can promote competition between new entrants and traditional utility incumbents.

The Regulatory Compact

The government response to the market imperfection of natural monopoly in the electricity industry was to regulate the industry. Ironically, regulation came in the form of a government-imposed monopoly. Simply, a private monopoly was replaced by a government-supported one. The government monopoly, through the regulatory compact, set prices at competitive levels, not above competitive levels, and did so through the process known as rate making, which is discussed herein.

Former Judge Kenneth Starr provides a good description of the regulatory compact:

> The utility business represents a compact of sorts; a monopoly on service in a particular geographical area (coupled with state-conferred rights of eminent domain or condemnation) is granted to the utility in exchange for a regime of intensive regulation, including price regulation, quite

> alien to the free market.... Each party to the compact gets something in the bargain. As a general rule, utility investors are provided a level of stability in earnings and value less likely to be attained in an unregulated or moderately regulated sector; in turn, ratepayers are afforded universal, non-discriminatory service and protection from monopolistic profits through political control over an economic enterprise.[18]

The compact, thus, is based on a quid pro quo. Privately owned utilities are subjected to government price and profit controls in exchange for which the utility undertakes a service obligation. Under the compact, consumers avoid monopolistic prices and are entitled to electric service. The utility is thus guaranteed a profit in an exclusive service area. The compact worked well for decades, as both producers and consumers benefited. Producers made a profit, and as utilities continued to enjoy economies of scale, consumer prices fell. In addition, public utility commission (PUC) cases were generally noncontroversial, and as a consequence, utility regulation fell below the political radar screen – that is, until prices rose precipitously in the 1970s.

Traditional Rate Making

Rate making is the device that drives the regulatory compact. The fundamental idea behind rate making is to mimic the market (i.e., to set prices at efficient or competitive levels). The primary objective is to enable a private utility to operate as a business. Rate making met other objectives as well.[19] Electricity was to be universally available, reliable, and abundant. Those objectives have been achieved as utilities built more and larger power plants and as they constructed a national electricity grid. In the early years of electricity regulation, as the industry expanded and prices declined, these goals were largely satisfied. Today, the continued use of the traditional formula comes at a cost that includes distorted market incentives and environmental pollution.

Rate Level

Traditional rate making was based on the simple, and then necessary, idea that a utility could satisfy its service obligation only if it received adequate revenue. The heart of the regulatory compact was the rate-making formula, which was designed to provide the utility with sufficient revenue to stay in operation. The formula can be simply expressed:

$$R = O + Br. \qquad (1)$$

Each of these variables is significant, as are the consequences on the industry from their application. *R* is the revenue requirement that is necessary for the utility to remain in business and serve its customers. The revenue requirement comprises operating expenses (*O*) and a rate base (*B*) on which the utility earned a rate of return. The rate base consisted of all of the capital investment made by the utility.

In essence, this formula is a type of cost plus contracting. A utility will recoup all of its prudently incurred expenses (*O*) and receive a government-determined rate of return (*r*) on its capital investment (*B*). The traditional rate-making process had the effect of encouraging capital investment because the larger the rate base, the greater the profits. Therefore, the rate formula rewarded utilities for building plants and selling electricity.[20] More is better. Because utilities were required to keep expenses prudent, they had the incentive to use the cheapest natural resource available to generate electricity (and profits), which has been and is coal.

Rate Structure

In addition to the rate level, electricity regulation is also driven by rate structure. For most of the twentieth century, the most prevalent structure was known as the declining block rate. The rate charged for the first period of consumption would be greater than in the second and subsequent periods. As a consequence, utilities were able to rely on high rates to recover their fixed costs in the first period and would recover their variable costs in subsequent periods. The price signal sent to consumers, however, indicated that the more energy they purchased, the cheaper that energy would be, thus encouraging consumption.

Consequences of Traditional Rate Making

The traditional rate-making formula and the regulatory compact contributed to distorted electricity markets. The formula led to excess capacity because utilities were rewarded for building more plants, and the compact led to high-cost electricity because utilities had monopoly service areas, which kept out new, cheaper providers. The regulations also contributed to environmental pollution because they encouraged coal use and discouraged the use of alternative and renewable energy resources. Having identified the problem with the old ways, the path to the new should be clearer. And it is.

Traditional utility regulation was based on the idea that bigger (and more) is better – the more electricity that was consumed, the healthier our economy would be. Thus, the guiding idea behind traditional electricity policy was to link energy growth to economic growth. That assumption must be replaced. Today, utilities must incorporate environmental health as well as economic health into their business plans, and they can do so by increasing energy efficiency and by improving the delivery of clean and efficient energy services and products. Regulators likewise must give utilities more operational flexibility in their operations. The ECS, by creating a race to the top, can help to move utilities toward incorporating environmental impacts into their bottom lines. The ECS would encourage utilities to employ cleaner technology by offering the incentive of collecting costs and a premium from dirtier companies.

Electricity and the Next Generation of Environmental Law

The challenges of global climate change warn us that unchecked electricity production causes environmental pollution and that our weak energy security makes us vulnerable in a contentious world. To answer these challenges, twenty-first-century energy policy must (1) use a diversity of natural resources for electricity production and distribution (2) in efficient and effective ways (3) while being environmentally responsible and (4) while promoting domestic and international security.

A new, renegotiated regulatory compact is an integral component of the moving beyond environmental law. A new regulatory compact will be based on three core ideas that also animate the Environmental Competition Statute. First, the ECS relies on competition and market-based regulations in place of command-and-control regulations. Second, the ECS recognizes that private competitive markets cannot provide public goods like a clean environment efficiently. Third, the ECS seeks to promote innovation, new technologies, and new markets.[21] All of these core ideas are congruent with the needed new regulatory compact. The ECS could be designed to work together with other elements of the compact to promote these ideas.

The New Regulatory Compact

Judge Starr's definition of the regulatory compact does not need to be reworded as much as it needs to be interpreted and applied in new ways. Under the new regulatory compact, the iUtility will sell energy services and products in a geographic market under a rate formula that gives the iUtility the necessary revenue to promote environmental protection while keeping rates just, reasonable, and nondiscriminatory.

The foundations on which the new regulatory compact should be built were laid as the industry changed in the 1970s and as regulators attempted to manage that change. Since the failed efforts at electric industry restructuring during the past two decades, electricity prices have risen above the anticipated rate because of the high cost of the resources used to produce electricity, especially natural gas and coal. Retail rates also increased as rate caps have been lifted. Additional pressure on prices has been caused by the need for additional generating plants and the great need for transmission investment. Finally, price pressure has also been brought about by increased concern for the environmental consequences of generation.[22]

As a result of increasing prices, utilities began to experiment with more environmentally responsible approaches by diversifying their fuel mix and including cleaner-burning resources as well as by engaging in conservation efforts. To address environmental concerns, utilities developed demand-side management (DSM)[23] and integrated resources planning (IRP). Both programs were intended to reduce demand during peak periods and to promote energy efficiency. Conservation, by reduced demand was to occur through consumers (DSM). The IRP programs were designed to encourage utilities to have a mix of resources that were more environmentally sensitive and efficient. Although these programs have not been touted as widely successful,[24] they were the beginning of a merger between energy regulation and environmental protection, and they are reappearing in regulatory proceedings in response to climate change, as state regulatory commissions begin to explicitly recognize the links among energy, the environment, and the economy.[25]

Regardless of these efforts, the electricity industry continues to face severe problems. The demand for electricity continues to grow; prices continue to rise; grid reliability is suspect; carbon emissions are unacceptable; and new entrants and new technologies are emerging, but entry and coordination into the old system remains troublesome. The response to these problems is the iUtility, which operates under a new regulatory compact.

iUtility Rate Making

The iUtility is distinguished from the traditional utility in that it will sell several ways. Most prominently, the iUtility will sell more varied products and services at more varied prices instead of selling electricity only. To accommodate the iUtility's revenue requirement, regulators must apply the rate formula more creatively, allowing the iUtility to recoup a more varied group of expenses and investments.

The iUtility and its regulation under the new regulatory compact will be designed to account for environmental investments by encouraging greater use of alternative and renewable resources, greater use of energy efficiency, and the development of new markets and new technologies. The ECS can work together with a new rate formula to facilitate the transfer of environmental investments from inefficient or dirty energy providers to more efficient and cleaner providers. The rate formula for the iUtility, like traditional rate making, comprises a rate level and a rate structure. Unlike traditional rate making, though, iUtility rate making employs each of these traditional components more expansively to achieve a broader set of clean-energy objectives.

Rate Level

The iUtility must have confidence in revenue stability; its customers must have confidence in rate stability; and regulators must have confidence that the rate formula achieves these and other social goals such as increasing energy efficiency and decreasing carbon emissions. Consequently, the new rate-making formula will enable the iUtility to earn its revenue requirement but O (operating expenses) and B (rate base) will be defined differently (see Equation 1).

The traditional formula encouraged both consumption and capital investment in more plants. The new formula contains different incentives, intended to encourage conservation, energy efficiency, and environmental protection rather than exclusively encouraging more investments in power production.

Rate Structure

The traditional utility was in the business of selling as much electricity as it could in order not only to meet its service obligation but also to earn its revenue and to reward its investors. For most of previous century, the rate structure was designed to encourage consumption through declining block rates. Utilities were able to recover their fixed costs early, and because rates were stable and on occasion declining, customers did not feel the pain of increasing prices. Those rates were also calculated on the basis of average costs, which do not accurately reflect the price of producing electricity, especially when the marginal cost of production exceeds average costs. More accurate price signals are derived from the marginal cost of producing the next unit of electricity.[26] The iUtility rates can be structured in one of several ways. Rates can be based on an increasing block structure in which the more electricity that is consumed the more expensive it becomes, thus encouraging conservation.[27] The iUtility can be required to charge customers on the basis of their marginal rather than average costs, or customers can be charged on the basis of either the time of use during the day or the season in which the electricity is consumed. Marginal cost pricing will inform customers about the varying cost of electricity, and consumers can then choose when and how much electricity to purchase at what prices.

The new rate making will enable iUtilities to earn revenue for their environmental investments and expenditures. The challenge for PUCs will be to determine (1) whether to treat expenditures as expenses, (2) whether to place expenditures into the rate base so that the utility can earn a return, (3) how to clearly define the nature of the product or service to be included in the rate formula, (4) how to determine for how long an iUtility can charge for the product or service, and (5) how to otherwise balance the interests of the iUtility for revenue against customer interests in reasonable rates against social interests in a cleaner environment.

Rate Design and the Environmental Competition Statute

The rate design for the iUtility can be made compatible with the goals of the ECS. To the extent that the ECS attempts to shift costs away from fossil fuel energy to cleaner alternative and renewable resources, the new rate design can take at least three forms. First, regulators can reward clean-energy projects with higher rates of return as an incentive to invest in such projects as solar and wind power, energy efficiency, and smart grid investments. Second, regulators can decide which type of investments should be included in the rate base and which should be treated as expenses. To the extent that investments are included in the rate base, the utility will earn a return on that investment instead of merely recouping its costs when an investment is treated as an expense item.

Third, and most important, the rate design can decouple energy sales from the utility's revenue requirement. Under traditional utility regulation, the utility earned its revenue based on the amount of electricity it sold. Consequently, the incentive was to sell as much electricity as possible. The new rate design will, instead, return revenue to the utility on the basis of a broad range of services and products, including the sale of energy efficiency. At its simplest, if the utility, through sales of services or products, reduces the customers' energy costs, those savings are shared between customer and utility. One form of such a rate design would involve allowing the utility to recover all of its fixed costs spread among all of its customers. In addition, a baseline of energy use per customer will be established. For those customers who exceed their baseline, they will pay an additional fee. Those customers who consumed less than the baseline are then given a rebate from the fees collected from the other users. The baseline then will be adjusted regularly to achieve efficiency and conservation goals. Under such a design, the utilities are indifferent relative to the charges made to customers because their fixed costs are recovered. Simultaneously, though, customers can conserve energy by consuming less than the baseline, and their rates will reflect that conservation.[28]

Renewable Portfolio Standards

The most promising regulatory reform that a PUC or legislature can adopt is known as a renewable portfolio standard (RPS). An RPS program can have dramatic consequences for changing traditional energy policy. To date, twenty-seven states and the District of Columbia use RPS programs to encourage the use of renewable and alternative energy resources, to stimulate new markets, and to support new technologies.[29] However, RPS programs have not fared as well in Congress. The Energy Independence and Security Act of 2007 failed to include a proposed provision for a 15-percent RPS requirement.[30] Collectively, the state policies apply to roughly 40 percent of the U.S. electricity load.

The idea behind RPS requirements is simple. The local utility is required to provide a specific percentage of its electricity from renewable energy sources such as wind, solar, or bioenergy. The obligation, then, is on the utility to purchase clean power in the market, thus reducing its dependence on fossil-fuel-generated electricity. The two key variables of these policies are the percentage of electricity that is to be distributed from renewable resources and the nomination of resources that satisfy the RPS requirement. Some states, such as Minnesota, have been very aggressive. Minnesota, for example, requires one utility to sell 30 percent of its electricity from renewable resources by 2020 and has placed a 25 percent requirement on all other electricity providers by 2025. By way of another example, Texas is expected to avoid 3.3 million tons of carbon dioxide emissions annually as it generates two thousand megawatts of renewable energy. To give some perspective on these requirements, it has been projected that an RPS of 20 percent by 2020 can have the effect of increasing total renewable energy capacity to 180,000 megawatts.[31] Throughout the states, RPS programs have enjoyed widespread bipartisan support. In addition, consumers and new producers have supported the policies. Properly designed, an iUtility will find RPS programs attractive because they are flexible and market based. Consumers will have a choice of fuels, and energy savings can also be made a part of an RPS program. States find that RPS programs support economic development as new technologies and new jobs become available.[32] The ECS seems to be a tool that could operate well in conjunction with an RPS program. An ECS seems to provide incentives for utilities to adopt cleaner technologies and to innovate to develop new energy sources. Such a statute should work well in parallel with an RPS program.

Technology Investments

No longer will the iUtility spend only to build plant; the iUtility will build plants and invest in a variety of technologies designed to produce energy and to increase energy savings. Consequently, investments in new technologies must be capable of being recovered in the iUtility's rates. Some of these technologies include smart meters,[33] advanced metering infrastructure,[34] and smart grid investment.[35]

Energy-Efficiency Surcharges

In addition to including expenses and capital investments in the rate formula, PUCs will consider adding an additional charge, an energy surcharge, onto customer bills to enable iUtilities to recover expenditures in energy-efficiency applications sometimes referred to as "negawatts." Negawatts equal the amount of energy saved. To the extent that a utility can assist with the installation of energy efficiency products or energy savings from complying with building codes, the utility can recapture those costs as either expenses or in the rate base. The state utility commission simply mandates that the utility has an obligation to invest in energy efficiency programming and then allows the utility to recoup that investment from its customers. As regulators impose energy performance standards for appliances, vehicles, and building, they must also determine how the iUtility's investment in satisfying those standards should be treated in their rates. The commissions could set spending budgets or savings targets such as through an energy-efficiency performance standard.

Disclosure

Public utility commission regulations can also require that the iUtility disclose its environmental and energy savings practices to customers and investors. Under a disclosure policy, the iUtility would be required to provide information regarding fuel sources or emissions associated with electricity generation and its energy investments. The key idea behind disclosure is to allow consumers to price and compare the resource mix and energy characteristics of their electricity purchases. Like green power pricing programs, disclosure is intended to enhance consumer choice. Choice is necessary for competitive and more efficient energy markets. Producers serving more than half of the electricity customers in the United States are subject to disclosure requirements, but it is not clear that disclosure requirements have stimulated new markets.[36] The core ideal behind disclosure requirements are to provide consumers with information, to increase choice, and to improve market efficiency. This requirement harmonizes well with the disclosure envisioned under the ECS.

iUtility Products and Services

The fundamental idea behind the iUtility is that the traditional utility will dramatically change its business model and, in addition to selling electricity, will offer a greater array of products and services. Of course, all electricity does the same thing, but the iUtility will offer electricity from traditional energy sources and from green sources, and the iUtility will offer to sell energy efficiency as a product. In addition, the iUtility will sell electricity at different prices at different times of the day or days of the year. The iUtility will offer pricing based on new rate-making formulas rather that the traditional of cost-of-service model. The iUtility may even impose a charge for environmentally responsive and energy-saving services. Included in the product mix will be pricing from local decentralized and distributed generation, thus reducing the scale at which electricity is produced. All power plants will not resemble giant tea kettles designed to boil water to turn turbines to generate electricity. In Amory Lovins's memorable phrase, "we do not need nuclear power plants to heat water, which is like cutting butter with a chainsaw."

iUtility Products

Electricity

The iUtility will stay with its core business and sell electricity. What changes is that the core business is supplemented by other products and services. Electricity will continue to cool our beer, run our computers, and let us watch football in high definition. Electricity will also play an expanded role in the transportation sector with electric and hybrid cars.

Green Electricity

Green power pricing is simply a requirement that utilities offer customers the option of buying green power. Consumers can choose to purchase an amount of power that would be generated from a renewable energy technology for which they will most likely pay the higher cost of electricity generated by renewable resources. In this way, green pricing programs can create markets for clean-energy tcchnologies.

Energy Efficiency

The importance of increased energy efficiency should not be underestimated. Energy efficiency may be the most economically efficient way of saving energy by reducing energy bills overall, reducing demand for fossil fuels, and stabilizing the energy system. Energy efficient gains in buildings, appliances, and cars are there to be made and in this regard, then, energy efficiency must be treated as a resource. Just as selling green power had its predecessors, so does selling energy efficiency.

Personal Energy Device

The iUtility can resemble its namesakes in one particular. Imagine a device, a personal energy profile and carbon footprint, which provides energy information. The device tells you the gas mileage on your car, the amount of energy lost in your home, the current prices of gasoline and electricity, alternative energy suppliers and products, energy-efficiency tips, and any other environmental and energy information you desire. The iUtility could provide the information, sell the device or the software, service the plan, and compile and synthesize customer information for its own business planning.

iUtility Services

The Efficiency Utility

Privately owned utilities will continue to play an important role in our energy future and therefore must remain financially healthy. Several states have actively engaged in efficiency planning, but the most advanced, forward-looking state has been Vermont.[37] Efficiency Vermont is a unique "utility" and is the first of its kind in the United States. It is a public utility charged with helping state residents save energy and protect the environment through energy-efficiency gains. It was created by the Vermont Public Service Board and is operated by the nonprofit service organization Vermont Energy Investment Corporation.

For several years, Vermont's electric utilities had been offering a range of energy-efficient services to their customers but had no great success to show for their effort. The programs were unconnected to one another and caused customer confusion. The idea was then born to coordinate energy-efficiency efforts under one roof. The efficiency utility is funded by an energy-efficiency charge (EEC) on consumers' electric bills. Early reports indicate that the EEC has caused little or no increase in monthly electricity bills for most customers. Efficiency Vermont provides technical assistance and financial incentives to customers to help reduce energy costs through energy-efficient equipment and lighting as well as energy-efficient approaches to construction and renovation.

Efficiency Vermont began operating in 2000 and has helped reduce energy costs in excess of $31 million. The program has also reduced summer peak loads and caused nearly a half-million-ton reduction in greenhouse gases over the lifetime of the installed measures. The report also notes that a net lifetime economic value for activities of 2006 could be in excess of $47 million, with total costs at roughly $28 million for a net benefit to the economy of $19 million.[38] The 2007 report states that the lifetime economic value is $76.1 million, at total costs of $39.7 million, with net lifetime benefits of $36.4 million.[39]

Energy Hedging

The core idea behind Enron, trading energy, was as brilliant as it is necessary. Unfortunately, greed, as we have seen, took over. Yet the collapse of the firm does not mean the collapse of the idea. The iUtility is not limited to selling electricity. Instead, it is an integrated energy provider. To survive, the iUtility must have a comprehensive understanding of its energy portfolio. What mix of energy resources, including negawatts, energy efficiency, energy futures, and carbon reduction strategies[40] must the utility have to (1) produce the electricity and other energy products and services that it will sell (2) at the lowest cost and (3) with the highest return?

The iUtility that is successful in designing such an energy portfolio will have developed expertise and will have created a valuable intellectual property that itself can be sold as a service. Part of this investment strategy can including hedging with energy and other energy-sensitive commodities or with investment-grade paper and interest swaps.[41] The iUtility will, then, become an energy trader and investment manager.[42]

Energy Audits

The iUtility that best manages its diversified energy portfolio and its energy investments will have accumulated another valuable intellectual property – energy advice. The iUtility can perform energy audits for its customers and then advise businesses, governments, and consumers about how best to realize energy savings, what energy mix is most valuable, how buildings can be constructed with the highest degree of energy efficiency, and which products are most efficient. In addition, the iUtility can advise those same customers about the range and options for putting together energy portfolios.

Conclusion

Traditional utility regulation has run its course. Over the past three decades, the electric utility industry and its regulators have been trying to redefine their relationship with little success. Climate change demands that we rethink fundamental assumptions about the relationship between energy and the economy; and climate change demands that energy production and consumption account for the high costs of carbon emissions.

Climate-change politics has its own strange bedfellows. New research organizations, new forms of venture philanthropy, new bipartisan political coalitions, and new nongovernmental energy organizations are proliferating. Most notably, there is an unanticipated actor on the climate-change stage – Wall Street. As the challenges of climate change mount, and as pressures on utilities continue, Wall Street is paying closer attention to investing in utilities with comprehensive programs for demand response, energy efficiency, and distributed generation.[43] More broadly, investment banks like Goldman Sachs and commercial banks like Citibank have sizable alternative and renewable energy investment divisions and are dedicating increasing amounts of capital for those investment opportunities. Venture-capital firms also see green investment as attractive. In 2007, more than $100 billion was invested in renewable energy projects, and that trend is anticipated to continue.[44]

To be sure, this new financial environment is anything but simple. Not so long ago, to calculate its investment needs, a utility needed only a straightedge to plot a 7-percent uphill graph and build accordingly, because growth in electricity demand was predictable and electricity was the only product being sold. Today, however, utilities and their investors must understand federal, state, and local taxes and regulations, as well as complex financial and economic evaluations; anticipate national and international politics; foresee technological innovations and their applications; and assess the costs, risks, and benefits of all of the foregoing. Today's utility executive, then, must:

> [U]nderstand the federal, state, and local tax incentives, and have the ability to make valuation adjustments to account for the legislative uncertainly associated with those benefits. In addition, an investor will require valuations of renewable energy credits produced or forfeited based on a generation-source choice for every applicable jurisdiction subject to a renewable portfolio standard. Further market expertise and prognostication will be necessary to incorporate the impact of carbon restrictions, as well as the impact these programs might have on the forward price of electricity. The ability to successfully draw upon each of these individual proficiencies, however, likely will not provide optimal efficiency in performing the overall economic analysis; to achieve a truly effective economic analysis, each of these individual considerations must be analyzed in an integrated and dynamic process.[45]

These risks can be reduced through a new regulatory compact between industry and government. At the heart of the compact will be a new rate formula designed to attract investors and bring stability to the regulatory environment. Such a new compact can be enhanced by the enactment of the ECS, which would create incentives for new and cleaner technology for electricity generation. Most specifically, the compact must (1) allow a return on investment by including energy efficiency investments in the rate base, (2) allow the recovery of fixed costs independent of volume throughput and electricity sales, and (3) contain regulations for cost recovery and return on investment that are clear and reliable and will serve the utility and its customers.

The iUtility will meet these challenges and will be poised to capture a good portion of new investment capital as it moves from an electricity-only provider to a provider of broad-based energy services and products. A forward-thinking power provider with technological savvy, a diversified energy portfolio, and a broad and varied mix of energy efficiency and environmentally sensitive products and services will emerge as the capital attraction winner among future energy producers. The new regulatory compact can meet the needs of climate change by transforming the stodgy old electric utility into the new iUtility, which delivers cleaner energy and promotes a healthier environment while sustaining economic growth.

NOTES

1. *See* GLOBAL CLIMATE CHANGE AND U.S. LAW 7–11 (Michael B. Gerard ed., 2007); *see also* Connecticut v. Am. Elec. Power, 406 F. Supp. 265, 268 (S.D.N.Y 2005).
2. NAT'L ENERGY POL'Y DEV. GROUP, NATIONAL ENERGY PLAN (May 2001).
3. Joseph P. Tomain, *To a Point*, 52 LOY. L. REV. 1201 (2006).
4. DEPT. OF ENERGY, ENERGY INFO. ADMIN., ANNUAL ENERGY REVIEW 2006 at 123 (June 2007).
5. *Id.* at 222.
6. *Id.* at 224.
7. GERRARD, *supra* note 2, at 9.
8. JOHN DEUTCH & ERNEST J. MONIZ (cochairs), THE FUTURE OF COAL: AN INTERDISCIPLINARY MIT STUDY 1 (2007). There are signs of coal prices increasing as U.S. exports increase. *See* Clifford Krauss, *U.S. Coal's Global Appeal: Exports Are Rising, and So Is the Price*. N.Y. TIMES, Mar. 19, 2008, at C1.
9. EDISON ELEC. INST., KEY FACTS ABOUT THE ELECTRIC POWER INDUSTRY (Feb. 2007).
10. *Id.* at 4.
11. *Id.* at 5.
12. LEONARD S. HYMAN, ANDREW S. HYMAN, & ROBERT C. HYMAN, AMERICA'S ELECTRIC UTILITIES: PAST, PRESENT AND FUTURE chs. 15–17 (8th ed. 2005).
13. *Id.* at ch. 18.
14. *See, e.g.*, Joseph P. Tomain, *The Past and Future of Electricity Regulation*, 32 ENVT'L L. 435 (2002); Sidney A. Shapiro & Joseph P. Tomain, *Rethinking Reform of Electricity Markets*, 40 WAKE FOREST L. REV. 497, 511 (2005).
15. *See, e.g.*, Jacqueline Lang Weaver, *Can Energy Markets Be Trusted? The Effect of the Rise and Fall of Enron on Energy Markets*, 4 Houston Bus. & Tax L.J. 1 (2004); NANCY RAPOPORT & JEFFREY D. VAN NIEL, ENRON AND OTHER CORPORATE FIASCOS: THE CORPORATE SCANDAL READER (2d ed. 2008); Darren Bush & Carrie Mayne, *In (Reluctant) Defense of Enron: Why Bad Regulation Is to Blame for California's Power Woes (or Why Antitrust Law Fails to Protect Against Market Power When the Market Rules Encourage Its Use)*, 83 OR. L. REV. 207 (2004).
16. U.S.-Canada Power Sys. Outage Task Force, *Final Report on the August 14, 2003 Blackout in the United States and Canada: Causes and Recommendations* (April 2004).
17. *Id.*
18. Jersey Cent. Power & Light Co. v. FERC, 810 F.2d 1168, 1189 (D.C. Cir. 1987).

19. *See* James C. Bonbright, Albert L. Danielsen & David R. Kameraschen, PRINCIPLES OF PUBLIC UTILITY RATES ch. 4 (2d ed. 1988); Charles F. Phillips Jr., THE REGULATION OF PUBLIC UTILITIES: THEORY AND PRACTICE ch. 5 (3d ed. 1993).
20. Harvey Averch & Leland L. Johnson, *Behavior of the Firm Under Regulatory Constraint*, 52 AM. ECON. REV. 1052 (1962).
21. WILLIAM J. BAUMOL, THE FREE-MARKET INNOVATION MACHINE: ANALYZING THE GROWTH MIRACLE OF CAPITALISM (2002); WILLIAM J. BAUMOL, ROBERT E. LITAN & CARL J. SCHRAMM, GOOD CAPITALISM, BAD CAPITALISM, AND THE ECONOMICS OF GROWTH AND PROSPERITY (2007).
22. Gregory Basheda et al., *Why Are Electricity Prices Increasing? An Industry-Wide Perspective*, 2–5 (June 2006).
23. The Department of Energy (DOE) reported on efforts under EPAct 2005 to look at demand response. Their report indicated that given the lack of uniformity in estimating and quantifying methods, that it was not possible to quantify the national benefits of demand response. Nevertheless, DOE recognized that demand response continued to have promise and recommended its continuation and studied consideration. DOE, *Benefits of Demand Response in Electricity Markets and Recommendations for Achieving Them: A Report to the United States Congress Pursuant to Section 1252 of the Energy Policy Act of 2005* (Feb. 2006). *See also* Hon. Jon Wellinghoff & David L. Morenoff, *Recognizing the Importance of Demand Response: The Second Half of the Wholesale Electric Market Equation,* 28 ENERGY L. REV. 389 (2007) (federal and state regulators should increase DSM requirements).
24. DOE Energy Info. Admin., *U.S. Electric Utility Demand-Side Management: Trends and Analysis*, *available at* http://www.eia.doe.gov/cneaf/pubs_html/feat_dsm/contents.html.
25. *Id.*
26. *See* Shapiro & Tomain, *supra* note 15.
27. *See* Ahmad Faruqui, *Inclining Toward Efficiency*, 146 PUB. UTIL. FORT. 22 (August 2008).
28. Ahmad Faruqui & Ryan Hledik, *Transition to Dynamic Pricing*, 147 PUB. UTIL. FORT. 26, 27–29 (March 2009); Ren Orens et al., *Inclining the Climate: GHG Reduction via Residential Ratemaking*, 147 PUB. UTIL. FORT. 41 (May 2009); *see* Wayne Shirley, Jim Lazar & Frederick Weston, *Revenue Decoupling: Standards and Criteria* (June 30, 2008), *available at* http://www.raponline.org/Pubs/MN-RAP_Decoupling_Rpt_6–2008.pdf.
29. Pew Ctr. on Global Climate Change, *States with Renewable Portfolio Standards*, *available at* http://www.pewclimate.org/what_s_being_done/in_the_states/rps.cfm (August 2007).
30. *See* FRED SISSINE, LYNN J. CUNNINGHAM & MARK GUREVITZ, ENERGY EFFICIENCY AND RENEWABLE ENERGY LEGISLATION 15–19 (2008).
31. Union of Concerned Scientists, *Experts Agree: Renewable Electricity Standards Are a Key Driver of New Renewable Energy Development*, available at http://www.ucsusa.org/clean_energy/clean_energy_policies.

32. *See* Apollo Alliance, *State Leadership for a New Energy Future: A Four-Point Initiative for Clean Energy and Good Jobs*, available at http://www.apolloalliance.org/downloads/resources_state_four_point_plan.pdf.
33. Smart meters will serve a range of functions, including facilitating net metering programs. In addition, smart meters will be needed to take advantage of marginal cost rates and time of use or real-time rates, and to more accurately monitor demand. Through better communication between consumers and the iUtility, smart meters can identify more accurate price signals, the demand on the grid, and signal available energy savings. Utilities can sell smart meters to customers and recoup the cost either as expenses or as a capital investment. Energy Policy Act of 2005, 16 U.S.C. §2621(d). Steve Lohr, *Monitoring Tools Help Cut Home Energy Use, Study Finds*, N.Y. Times, Jan. 10, 2008, at C1; *see also* two reports for the DOE's Pacific Northwest National Laboratory: D.J. Hammerstrom et al., *Pacific Northwest GridWise Testbed Demonstrations Project Part I. Olympic Peninsula Project* (October 2007), and D.J. Hammerstrom et al., *Pacific Northwest GridWise Testbed Demonstrations Project Part II. Grid Friendly Appliance Project* (October 2007).
34. Advanced Metering Infrastructure (AMI) provides the iUtility information infrastructure for billing, assessing load and demand, and evaluating grid reliability and security. *See generally* Plexus Res., IMC, *Deciding on "Smart" Meters: The Technology Implications of Section 1252 of the Energy Policy Act of 2005* (Sept. 2006). The good news on AMI is that the technology is available. The challenge is that it has been difficult implementing these technologies so that smart meters and AMI can be properly integrated into a utility's customer information and billing services at the user end of the iUtility's business and into transmission and distribution systems at the other. Michael T. Burr, *Smart Grid, Smart Utility*, 145 Pub. Util. Fort. 39 (Feb. 2007).
35. The Northeast's blackout of 2003 was a wake-up call for grid reliability. Estimates vary, but there is a consensus that IOUs have not invested to the extent that they should have to keep the grid up and reliable. *See* Roger Wood, *Banking on the Big Build*, 145 Pub. Util. Fort. 49 (Oct. 2007). Most notably, investments must be made in a more technologically sophisticated grid, not in the same old set of high-voltage transmission lines. To that end, to the DOE's Office of Electricity Delivery and Energy Reliability recently announced more than $50 million of funding to modernize the U.S. electric grid.
36. Richard P. Sedano, Electric Product Disclosure: A Status Report (July 2002).
37. Efficiency Vermont, 2006: Preliminary Executive Summary; Efficiency Vermont, 2006: Preliminary Results in Saving Estimates Report (Mar. 30, 2007).
38. *Id.*
39. Efficiency Vermont, 2007 Annual Report: Executive Summary (Oct. 15, 2009).
40. Revis James et al., *The Power to Reduce CO_2*, 145 Pub. Util. Fort. 60 (Oct. 2007).
41. Stephen Maloney, *When the Price Is Right*, 145 Pub. Util. Fort. 24 (Oct. 2007).

42. Terry Pratt et al., *Rating the New Risks: How Trading Hazards Affect Enterprise Risk Management at Utilities*, 145 PUB. UTIL. FORT. 28 (June 2007); *see also* Timothy P. Gardner & James C. Hendrickson, *Carbon Wargames: U.S. Utilities Gain Strategic Insights by Playing Out a Carbon-Constraint Scenario*, 145 PUB. UTIL. FORT. 46 (Dec. 2007).
43. Tom Brunetto, *The Color of Money*, 145 PUB. UTIL. FORT. 19 (Oct. 2007).
44. *See* REN21, *Renewables* 2007: GLOBAL STATUS REPORT 6 (2008), *available at* http://www.ren21.net/pdf/RE2007_Global_Status_Report.pdf.
45. Elias Hinckley, *Before You Build It: Think Green*, 145 PUB. UTIL. FORT. 54, 57 (Oct. 2007).

11 Environmental Patriotism

Christine A. Klein

THIS BOOK IS ABOUT THE FUTURE OF ENVIRONMENTAL LAW. To understand our proposals for the next generation of environmental law, it is useful to look backward to the first two generations of environmental law. Each stage grew out of quintessential American values, articulated at the founding of our nation but equally applicable to subsequent generations. Historically, the most successful laws have codified shared values rather than imposed the will of a few on the many (as did the unsuccessful Prohibition laws). Overall, our most enduring legal traditions can be viewed as an expression of patriotism. As perhaps the core American value, patriotism is a rich, fertile soil containing essential nutrients – liberty, equality, happiness, optimism, strength, frugality, efficiency, industriousness, and individualism, to name but a few. This soil nourishes a variety of cultural norms and social institutions, many of which ripen into law. Just as different species of trees may thrive in different soil types – river birch in alluvial soils, the giant sequoia in deep sandy loams, the black walnut in well-drained clays – so also vigorous legal regimes draw on a mix of values well suited to the context.

Drawing on these durable values, this book charts a course forward, encapsulated in two proposed legal frameworks: the National Environmental Legacy Act and an Environmental Competition Statute. Although pathbreaking in orientation, the proposals are rooted firmly in basic American values. As such, they fit comfortably within the tradition this chapter calls environmental patriotism – the belief that a healthy natural environment

I am grateful to Ling-Yee Huang and Samantha Alves Orender for their research assistance.

promotes the strength, safety, and well-being of the nation. This linkage of environmental laws and patriotism suggests that the priority placed on environmental protection deserves to be on a par with that of national security.

This chapter surveys America's unique brand of patriotism, tracing its deep roots in the country's physical landscape. The chapter tracks our evolving sense of patriotism, as expressed through art, literature, and law. It then describes how environmental law incorporated patriotic values throughout its first and second generations, focusing on the key values of optimism, strength, and thrift. Finally, the discussion introduces this book's proposals for the next generation of environmental laws, noting how they continue the American tradition of environmental patriotism.

Patriotism and the Land

In the United States, patriotism is strongly rooted in the land. Writing in the early years of the Republic, Thomas Jefferson identified a fundamental link between democracy and the small family farm. He asserted, "Cultivators of the earth are the most valuable citizens. They are the most vigorous, the most independent, the most virtuous, and they are tied to their country and wedded to its liberty and interest by the most lasting bands."[1] Jefferson noted with pride the uniqueness of the American circumstance, particularly its "immensity of land." Such natural abundance, Jefferson believed, could support a nation of landowning farmers, freed from the dependence upon others that "begets subservience and venality, [and] suffocates the germ of virtue."[2]

This special connection between patriotism and the land is a hallmark of American society that has significantly influenced the development of both art and law. In the nation's early years, landscape painters rejected the rough, untamed wilderness as a valid focus for their art. At times, the wild landscape was even perceived as a national embarrassment, comparing unfavorably with the refined European countryside. One art historian explained, "Pure landscapes were unheard of since portraits and religious or historical scenes were considered the only valid subjects for art in the new world."[3] Despite this early view, by about 1825, America's first homegrown school of painting – the Hudson River School – began to emerge from a group of artists active in the New York area. Influenced by writers such as Henry David Thoreau and James Fenimore Cooper, the artists produced dramatic canvases depicting luminous, largely untouched, identifiably American wilderness. As one museum director explained, "[This was a time in American history] when the country was grappling with issues of national identity." The Hudson River paintings "show a country distinguishing itself from Europe and celebrating what was uniquely American – the landscape."[4]

American landscape art directly influenced the development of conservation law. One of the Hudson River painters, Thomas Moran, traveled with the F. V. Hayden expedition as it explored the Yellowstone Valley of Wyoming and Montana. The expedition's *Comprehensive Report* – prominently featuring Moran's artwork – was credited with influencing Congress to set aside Yellowstone as a protected area, holding it in a perpetual federal trust as a legacy for all Americans. In 1872, toward the end of the Hudson River school era, President Ulysses S. Grant signed legislation making Yellowstone the world's first national park, dedicating it as a "pleasuring ground for the benefit and enjoyment of the people."[5]

Thus, America's emerging environmental patriotism – its embrace of the rugged natural landscape as central to national identity and well-being – found expression through both art and law. The environmental historian William Cronon explained the connection between the preservation of natural areas and fundamental American values: "[A]lthough today we protect wild places to preserve biological diversity and other natural values, we should never forget how deeply tied they are to American ideas of God and nationhood." In particular, "[A]s a nation born of the Romantic era, coming to full national identity in the 19th century, we saw in our most wild and beautiful natural lands symbols of the sublime – places where the divine was most present in the world."[6] In modern times, our national identity remains rooted in the land, even as the landscape becomes increasingly urbanized.

The First Generation of Environmental Law (c. 1960–80)

American environmental law began to come of age in the 1960s. In 1962, Rachel Carson published her influential book *Silent Spring*, bringing the nation's attention to the environmental impacts of the pesticide DDT. Congress passed a spate of federal environmental statutes during this period, beginning with the Clean Air Act of 1963. Milestones include the Air Quality Act of 1967; the National Environmental Policy Act of 1969;

the Clean Air Act Amendments of 1970; the Federal Insecticide, Fungicide, and Rodenticide Act of 1972; the Federal Water Pollution Control Act of 1972; the Marine Protection, Research, and Sanctuaries Act of 1972; the Endangered Species Act of 1973; the Safe Drinking Water Act of 1974; the Resource Conservation and Recovery Act of 1976; the Toxic Substances Control Act of 1976; the Clean Air Act Amendments of 1977; the Clean Water Act Amendments of 1977; and the Comprehensive Environmental Response, Compensation, and Liability Act of 1980. This legislation drew on numerous American values, including optimism, strength, and thrift.

Optimism

In the aftermath of World War II, Americans looked forward with optimism, enjoying the new prosperity that rewarded their wartime sacrifice. The baby boom continued until about 1965, reflecting Americans' confidence that their children – and their children's children – would have ever-more opportunities for a bright future. As the first generation with widespread access to television, the children of this boom period saw an idealized version of the white American family reflected in shows such as *Father Knows Best* and *Leave It to Beaver*. Echoing Thomas Jefferson's reverence for America as the "chosen country, with room enough for our descendants to the thousandth and thousandth generation,"[7] the postwar families flocked to new suburban subdivisions, finding "[a] quiet place where yards are wide, people few, and motor vehicles restricted."[8]

During this time, Dr. Martin Luther King Jr. expressed his optimistic vision for racial equality through land-based metaphors. In his famous "I Have a Dream" speech of 1963, King spoke of the day when "all of God's children will be able to sing with a new meaning, 'My country, 'tis of thee, sweet land of liberty, of thee I sing.' " Drawing from the song's refrain, he asserted: "So let freedom ring from the prodigious hilltops of New Hampshire.... Let freedom ring from the curvaceous peaks of California! ... From every mountainside, let freedom ring."[9]

The first generation of environmental laws reflected this forward-looking optimism. The Clean Water Act, for example, set forth the wildly ambitious goals of achieving fishable and swimmable waters by 1983[10] and of entirely eliminating water pollution by 1985.[11] Numerous other statutes of the period were similarly forward looking, recognizing a duty to maintain a legacy for future generations. The National Environmental Policy Act, for example, created a federal obligation to take all practicable measures to "fulfill the responsibilities of each generation as trustee of the environment for succeeding generations"[12] and to "preserve important historic, cultural, and natural aspects of our national heritage."[13]

Strength

During this era, the country focused on the values of national strength and individual health. At the national level, shows of strength played an important role in the United States' Cold War against the Soviet Union and in U.S. efforts to contain communism. The two nations engaged in a competitive space race couched in patriotic terms. After the Soviet Union sent the first human into space on April 12, 1961, the United States followed closely behind through its *Freedom Seven* mission of May 5, 1961. Later that month, President John F. Kennedy set landing a human on the moon as a national goal, describing it as an essential achievement "[i]f we are to win the battle that is now going on around the world between freedom and tyranny." Kennedy's call came to fruition on July 21, 1969, when the *Apollo 11* mission's Neil Armstrong and Buzz Aldrin became the first people to walk on the moon. The patriotic space mission intersected with the environmental movement in 1972, when the *Apollo 17* crew took the famous "blue-marble" photograph of Earth from about eighteen thousand miles away, a photo that became emblematic of the need to protect our beautiful, fragile, and isolated planet.

Complementing the focus on national strength, the country also sought to ensure the strength and health of its individual citizens, particularly children. In 1956, President Dwight D. Eisenhower founded the President's Council on Physical Fitness and Sports (originally, the President's Council on Youth Fitness), after he was reportedly "shocked" and "alarm[ed]" by a study published in the *New York State Journal of Medicine* that found American children to be decidedly weaker and less fit than their European counterparts, a problem attributed to an affluent, sedentary lifestyle.[14] President Kennedy took up the charge. In addition to a massive public education campaign, Kennedy's council promoted exercise routines for school gym classes. Some readers may recall the whimsical "Chicken Fat Song," composed for Kennedy's council, which urged schoolchildren to perform sit-ups, push-ups, and toe touches in time with its energetic beat.[15] These were also the Wonder bread years, during which the Continental Baking Company boasted that its vitamin- and mineral-enriched product "helps build strong bodies 12 ways."[16] During this time, President Lyndon B. Johnson drew an explicit link between national strength and individual health: "The fitness of our nation for the tasks of our times can never be greater than the general physical fitness of our citizens. A people proud of their collective heritage will take pride in their individual health, because we cannot stay strong as a country if we go soft as citizens."[17]

The national premium placed on physical health, strength, and fitness influenced the environmental legislation enacted during that period. Health-based standards predominated. The Clean Air Act, for example, required the Environmental Protection Agency to set national "primary ambient air quality standards" at levels requisite "to protect the public health [allowing an adequate margin of safety]."[18] Similarly, the Clean Water Act's water quality standards sought, in part, to "assure protection of public health."[19] Other statutes also recognized health goals, such as the National Environmental Policy Act's goal to "assure for all Americans safe, healthful, productive, and esthetically and culturally pleasing surroundings."[20]

Thrift

Americans are frugal by nature. Historically, we have demonstrated a desire to conserve and an aversion to waste, even during periods of plenty. In 1780, future senator and congressman Jonathan Mason warned, "Remember that prosperity is dangerous.... [B]ut while patriotism is the leading principle,... while industry, frugality, and temperance are held in estimation,... peace and affluence will throw their smiles upon the brow of the individual, [and] our commonwealth will flourish."[21] Later, the American transcendental movement extolled the virtues of economy, frugality, and a simple life in harmony with nature, as expressed by writers including Walt Whitman, Ralph Waldo Emerson, and Henry David Thoreau. During periods of war and international strife, frugality became even more important. For example, World War II poster art encouraged conservation measures such as carpooling. In a particularly grim series of posters, a battle-hardened GI asked, "Have you *really* tried to save gas by getting into a car club?"[22] In a similar vein, another poster depicted a lone driver, seemingly oblivious to the shadowy figure of Adolf Hitler seated beside him in the passenger seat. The caption admonished, "When you ride *alone* you ride with Hitler! Join a car-sharing club *today!*"[23]

The first generation of environmental law codified the virtue of thrift, generally through the rhetoric of conservation and sustainability. The Multiple-Use Sustained-Yield Act of 1960, for example, directed the secretary of agriculture to administer the national forests for "sustained yield," defined as "the achievement and maintenance in perpetuity of a high-level annual or regular periodic output of the various renewable resources of the national forests without impairment of the productivity of the land."[24] Slightly more than a decade later, in 1973, thrift in the form of conservation took center stage when the Arab nations placed an embargo against the export of crude oil to the United States and other Western countries. Gas prices rose from an average of 38.5 cents per gallon in May 1973 to a high of 55.1 cents per gallon in June 1974. In response, the federal government enacted a number of conservation measures, including the Emergency Highway Energy Conservation Act of 1974 (capping speed limits nationwide at fifty-five miles per hour), the Corporate Average Fuel Economic (CAFE) standards of 1975, and the National Energy Act of 1978. The idea of conservation also figured prominently in the Endangered Species Act of 1973, designed, among other things, to "provide a means whereby the ecosystems upon which endangered species and threatened species depend may be conserved."[25]

The Second Generation of Environmental Law (c. 1981–2008)

As Ronald Reagan swept into the Oval Office in 1981, he embraced the long-standing values of optimism, strength, and thrift. But whereas the first environmental generation relied on a strong federal government to protect, cajole, and inspire, President Reagan and some of his successors rejected such efforts. Instead, they placed their faith in the resourcefulness of the individual and in the free market. The second generation of environmental laws experimented with the free market and produced some important successes that can serve as models for the future. But the second generation also dismantled important environmental safeguards, teaching us that, without sufficient government oversight, the free market can run amok, violating cherished values of optimism, strength, and thrift.

Optimism

President Reagan ushered in a period of sunny optimism, as captured by his campaign slogan, "It's morning again in America." In the 1984 campaign leading to his election to a second presidential term, Reagan aired a television spot that proclaimed:

> It's morning again in America. Today more men and women will go to work than ever before in our country's history. With interest rates at about half the record highs of 1980, nearly 2,000 families today will buy new homes, more than at any time in the past four years. This afternoon 6,500 young men and women will be married, and with inflation at less than half of what it was just four years ago, they can look forward with confidence to the future. It's morning again in America, and under the leadership of President Reagan, our country is prouder and stronger and better. Why would we ever want to return to where we were less than four short years ago?[26]

Notably, Reagan tied his optimism to the perception that government deregulation had led to economic prosperity, a proposition rendered debatable by the economic recession that would be declared in 2008. Nevertheless, as explained herein, Reagan continued his aggressive deregulatory initiatives, weakening environmental protections and government oversight.

Strength

The second generation of environmental laws honored strength but focused on the individual (as strong, creative entrepreneur) rather than on the collective (as strong, protective government). For example, President Reagan trusted the "creative energy" and "individual genius of man," casting ordinary citizens as patriotic heroes: "You can see heroes every day.... They are entrepreneurs with faith in themselves and faith in an idea who create new jobs, new wealth and opportunity.... Their patriotism is quiet but deep. Their values sustain our national life."[27] The president envisioned an important but limited role for the federal government, seeking to make it "work with us, not over us; to stand by our side, not ride on our back. Government can and must provide opportunity, not smother it; foster productivity, not stifle it." Consistent with the focus on market efficiency, and the lack of attention to market failures in this period, President Reagan aggressively pursued a policy of deregulation – simplifying and streamlining government mandates – thereby freeing the market from what he perceived to be excessively onerous and inefficient government oversight.

Second-generation environmental laws bear the mark of this reliance on the free market and the strength of individual innovation. For example, the Clean Air Act Amendments of 1990[28] established a market-based cap-and-trade program for the reduction of acid rain. As the Environmental Protection Agency (EPA) describes this scheme, each relevant polluting facility was issued a certain number of allowances, each giving the right to emit one ton of sulfur dioxide pollution into the atmosphere:

> If a plant expects to release more SO_2 than it has allowances, it has to purchase more allowances or use technology and other methods to control emissions. A plant can buy allowances from another power plant that has more allowances than it needs to cover its emissions. There is an allowances market that operates like the stock market, in which brokers or anyone who wants to take part in buying or selling allowances can participate. Allowances are traded and sold nationwide.[29]

This program – rewarding resourceful technological innovation and allowing individual actors greater control and flexibility – has been recognized as "perhaps the greatest green success story of the past decade" by the *Economist.*[30] Likewise, a prominent environmental group acknowledged the ability of properly harnessed market forces to achieve "superior economic protection by giving businesses flexibility and a direct financial incentive to find faster, cheaper and more innovative ways to reduce pollution."[31] When improperly harnessed, however, market forces fall short of protecting the environment.[32]

Thrift

During the second environmental generation, the time-honored value of thrift was recast as efficiency. In the technological realm, this was indeed the era of efficiency. Thanks to the development of the microprocessor and the silicon chip in the 1970s, a room full of computer equipment could now fit easily on the top of a desk. In 1981, the first IBM home computers hit the market, and in 1983, Compaq released its first "portable" computer – weighing in at a hefty twenty-eight pounds. In business, 1980 was the decade of corporate takeovers, and the 1990s were the time of corporate downsizing – all in the name of efficiency.

Government, too, focused on efficiency. President Reagan sought to create a leaner, more efficient government. In his first inaugural address, Reagan asserted, "In this present [economic] crisis, government is not the solution to our problem," and he stated his intention to "curb the size and influence of the Federal establishment." Implementing so-called Reaganomics, the president attacked the nation's high rates of inflation, unemployment, and deficit spending with dramatic tax and spending cuts. Little more than a decade later, the Republican Party would continue its push for government efficiency through its "Contract with America," calling for, among other things, measures to stop governmental waste, fraud, and abuse, and to reduce the influence of government altogether. History has rendered a mixed verdict as to the success of these policies. Reliance on the free market was seemingly vindicated by two contrasting developments during the 1990s. Communism failed, as witnessed by the collapse of the Soviet Union in 1991.[33] At the same time, capitalism flourished in the United States, which enjoyed sustained economic expansion and a booming stock market during the second half of the 1990s. And yet troubling signs also appeared. In the absence of federal regulation,[34] some thousand savings and loan institutions failed during the 1980s, bailed out by a $125 billion federal subsidy. By century's end, the dot-com bubble burst, as technology stocks sharply declined in value. By 2006, another financial crisis developed, this time in the subprime mortgage market. Foreclosures rose sharply, attributable in part to the lack of adequate oversight in the lending market.[35] In 2008, the federal government announced its takeover of the giant mortgage-finance institutions Fannie Mae and Freddie Mac, and it acknowledged that the economy was officially in recession. By the time the Obama administration took office, a newfound recognition of the need to regulate markets supplanted the gospel of the free market and refocused attention on many instances of market failure that justify rigorous governmental oversight.

This focus on efficiency found its way into the law. Early in his first term, Ronald Reagan set the tone for his presidency with Executive Order

No. 12,291, which required the preparation of "regulatory impact analyses" for all major rules, and precluded regulatory action "unless the potential benefits to society for the regulation outweigh the potential costs to society."[36] This order had significant ramifications for environmental protection, requiring regulators to calculate the worth of potentially priceless extramarket environmental values in dollar terms. On a more constructive note, legislation of the time emphasized the sharing of information, thereby enhancing the availability of knowledge thought essential to the efficient functioning of markets. Information-related statutes included the Emergency Planning and Community Right-to-Know Act of 1986[37] and the Regulatory Right-to-Know Act of 2001.[38] Under the latter, the Office of Management and Budget (OMB) must prepare an annual report estimating the total costs and benefits of major federal rules imposed by specified federal departments and agencies. Belying the period's distrust of regulation, the OMB's 2005 report found that major regulations imposed by the EPA between 1994 and 2004 generated annual benefits of at least $44.3 billion, an amount triple the compliance costs that they exacted.[39]

The Next Generation of Environmental Law (c. 2009–?)

The next generation of environmental law must draw on the value of strength, combining the powerful federal leadership of the first generation with the strong, individual initiatives of the second generation. It must be optimistic in reach, and it should strive to thriftily conserve our natural resources in the most efficient way possible. Americans clearly want a robust federal government that is able to deliver essential services. They have bitterly noted the failed leadership of numerous federal agencies, including the Department of Transportation's failure to adequately maintain the Interstate 35 bridge that collapsed in Minneapolis in 2007; the Food and Drug Administration's failure to prevent the import of poisoned toothpaste, *E. coli*–tainted spinach, toxic pet food, and lead-painted toys into American markets; the Department of Veterans Affairs' failure to prevent the deterioration of Walter Reed Hospital; and the Federal Emergency Management Agency's failure to conduct an adequate operation in New Orleans during the devastating 2005 Atlantic storm season.

At the same time, Americans fear excessive government power and want federal action that is clearly focused, even surgically targeted. As David

Brooks of the *New York Times* explains, "[Voters] want a federal government that will focus on a few macro threats – terrorism, health care costs, energy, entitlement debt and immigration.... This is not liberalism, which inserts itself into the crannies of life. It's not conservatism, suspicious of federal power. It's a gimlet-eyed federalism – strong government with sharply defined tasks."[40] In the environmental realm, some of America's corporate leaders took up the charge in a move that would have been largely unthinkable just a decade earlier. The U.S. Climate Action Partnership – with members including Alcoa, British Petroleum America, Caterpillar, Duke Energy, DuPont, General Electric, Lehman Brothers, and PG&E Corporation – demanded that the federal government quickly enact strong national legislation requiring significant reductions of greenhouse gas emissions, thereby providing a federal baseline from which innovators can move forward.[41]

Beginning in about 2006, the country set the stage for the development of a new generation of environmental law. In 2006, the Democratic Party made its strongest showing since 1994, with the Democrats taking over a majority of seats in the House of Representatives (233–202), in the Senate (51–49), and a majority of state governorships. The new speaker of the House – and the first female speaker ever – Nancy Pelosi (D-California), vowed to take action on climate change and on national energy policy. With the election of President Barack Obama and Democratic control of Congress confirmed in November 2008, and with widespread recognition of the reality and scale of the threat posed by climate change, prospects for a new generation of environmental law began to emerge, emphasizing the connection between patriotism and environmental conservation.

Green Is the New Red, White, and Blue

The first generation of environmental law relied heavily on federal command and control, with the federal government issuing numerous commands to regulated industries and controlling, to a large extent, the methods by which those commands could be satisfied. During the second generation of environmental law, the pendulum swung dramatically in the opposite direction, as political leaders sought to reduce both the size of the federal government and the degree of federal oversight and regulation. Both generations found sustenance in deeply rooted patriotic values. But the translation of those values into concrete action, as with most things, revealed both strengths and weaknesses of the relevant legislation. The next generation of environmental law should take advantage of the best of each of the preceding two generations and avoid the mistakes of the past.

The fundamental importance of environmental law has become increasingly apparent during its relatively brief history. At times, environmental values have been treated as secondary to the nation's core interests, dismissed as a luxury that only the elite can afford. Some have gone farther and cast environmental protection as an impediment to economic growth and full employment rates.

Early in the twenty-first century, however, a paradigm shift began to emerge, accelerated by a trio of events. First, the terrorist attacks of September 11, 2001 – together with the wars in Afghanistan and Iraq – made painfully apparent the danger to national security of relying on foreign oil. To create a more durable national energy policy, environmental conservation will undoubtedly figure into the mix, including the transition to renewable, sustainable sources of energy. Second, the record-breaking Atlantic storm season of 2005, including the devastation wrought by Hurricanes Katrina and Rita, caused many to recognize the importance of a strong federal government in times of need and caused many to wonder what had gone awry with weather patterns in the nation and in the world. Third, the issue of global warming saturated the media soon thereafter, following the 2006 release of Al Gore's documentary, *An Inconvenient Truth*; the 2007 U.S. Supreme Court decision in *Massachusetts v. EPA* (holding that the EPA has statutory authority under the Clean Air Act to regulate the emission of greenhouse gases by new motor vehicles); and the 2007 release of *Climate Change 2007*, a series of reports by the Intergovernmental Panel on Climate Change (IPCC), established in 1988 by the World Meteorological Organization and the UN Environment Programme. Among other things, the IPCC reports addressed the connection between climate change (including global warming) and weather patterns, predicting for the twenty-first century an average surface temperature increase of 1.8 to 4.0 degrees centigrade and global mean sea-level rise of 0.18 to 0.59 meters. The IPCC also found a greater than 66 percent likelihood that the intensity of tropical hurricanes and typhoons would increase over the coming century.

Following this trio of events, environmental protection was increasingly identified as fundamental to national security and in sync with cherished values.[42] In 2007, Congress held a hearing on *Climate Change: National Security Threat* before the Committee on Foreign Relations. At that hearing, the Republican Senator Richard Lugar asserted, "[W]e need to understand how climate change might be a source of war and instability," and then Democratic Senator Joseph Biden testified, "Climate, energy, national security – these are all facets of the same single challenge." President George W. Bush acknowledged the value of environmentalism, at least in theory: "The federal government has . . . responsibilities . . . to protect the environment. America's experience with environmental protection really makes me optimistic about our future. The modern environmental movement was born in America, born right here in our country."[43] President Obama made environmental sustainability a prominent centerpiece of his administration, stating during his presidential campaign, "For the sake of our economy, our security, and the future of our planet, I will set a clear goal as president: In 10 years, we will finally end our dependence on oil from the Middle East." Similarly, in his inaugural speech, President Obama observed, "Our challenges may be new, the instruments with which we meet them may be new, but those values upon which our success depends, honesty and hard work, courage and fair play, tolerance and curiosity, loyalty and patriotism – these things are old."[44]

Furthermore, the national discussion began to link environmental protection with a bipartisan sense of patriotism. The *New York Times* columnist Thomas Friedman wrote extensively on the "power of green." Friedman explained, "I want to rename *green.* I want to rename it geostrategic, geoeconomic, capitalistic, and patriotic. . . . A redefined, broader and more muscular green ideology is not meant to trump the traditional Republican and Democratic agendas but rather to bridge them when it comes to addressing the three major issues facing every American today: jobs, temperature and terrorism."[45] Friedman posited that "living, working, designing, manufacturing and projecting America in a green way can be the basis of a new unifying political movement for the 21st century."[46] A number of groups joined Friedman in touting green as the new red, white, and blue. The Civil Society Institute, for example, asserted, "Red, white and green is the new patriotism for this complex era in which energy, the environment and national security are all bound up together. What we see here are a new set of values that equate love of country with energy independence and protecting the environment from global warming."[47] According to a 2005 survey conducted by the Institute, two-thirds of Americans believed that the purchase of fuel-efficient vehicles was patriotic because it made the nation less dependent on oil imports from the Middle East.[48]

The National Environmental Legacy Act

Drawing on the lessons of history, the National Environmental Legacy Act proposed in the first part of this book assigns the federal government a focused task eminently suited to its great stature and consistent with core American values: serving as a guardian for the interests of future generations of unborn Americans with respect to the nation's stock of federally owned lands. This is an optimistic task – taking concrete steps to ensure that our children will enjoy a homeland as wondrous and nurturing as the one that sustained the current generation. Under the Legacy Act, the federal government would establish a long-range legacy period for planning, thereby transcending the market's failure to look beyond short-term planning cycles that are better calibrated to the preparation of shareholder quarterly statements than to running the affairs of a great nation. The legislation also promotes strength: it draws on health-based standards and recognizes that the degradation of natural resources, including a healthy atmosphere free of greenhouse gases, robs the country of its national strength and weakens the health of its citizens. Finally, the proposed legislation is also an expression of thrift, promoting fiscal responsibility by preventing the current generation from spending down the public's stock of natural resources, assets that belong also to future generations.

An Environmental Competition Statute

The next generation of environmental law must not rely on government alone but must also draw from the strength of individual ideas and innovation. Modern-day environmental patriots have already begun to emerge, from government and elsewhere. In 2007, Al Gore and the IPCC jointly won the Nobel Peace Prize for their "focus on the processes and decisions that appear to be necessary to protect the world's future climate, and thereby to reduce the threat to the security of mankind."[49] Banking on entrepreneurial solutions to environmental problems, in 2007, the British billionaire Richard Branson offered a prize of $25 million to the first person able to remove at least 1 billion tons of carbon dioxide annually from the atmosphere.[50] Religious groups have also called for individual action, framing environmental protection as a moral imperative. The Interfaith Center on Corporate Responsibility (ICCR), for example, called on its 275 faith-based institutional investors to sponsor shareholder resolutions on major social and environmental issues.[51] The ICCR's working group on global warming states as its mission: "to educate companies about the environmental and economic threats posed by the emissions of greenhouse gases from their products and operations, and to increase shareholder value by urging companies to proactively address the global warming challenge."[52] Perhaps one of the most colorful campaigns – What Would Jesus Drive? – asserted that "transportation is a moral issue" and promoted "new ways to love your neighbor [by striving] together to reduce fuel consumption and pollution from the cars, trucks, and SUVs we drive."[53]

The Environmental Competition Statute proposed in the second part of this book relies on the creativity of citizens and businesses to develop the mechanisms for a better environmental future. Authorizing low-polluting entities to recoup their pollution reduction costs from higher-polluting entities, the proposed statute provides economic incentives for a race to the top through the development of new, environmentally sustainable technologies. The statute draws on the American value of optimism, believing that we can invent new solutions to cut our emissions of greenhouse gases and other pollutants. In addition, the statute relies on the strength of the federal government to stimulate individual resourcefulness. Furthermore, the statute calls for thrift, creating a highly competitive environmental market engaged in a race to produce cost-effective and environmentally superior technology.

The two statutes that form the backbone of this book look backward to the essential values that have supported our nation and look forward to future challenges that we cannot yet imagine. We hope that our proposals stimulate the imagination of our readers to come up with yet more third-generation environmental protections.

NOTES

1. Thomas Jefferson, *Letter to John Jay*, Aug. 23, 1785.
2. Thomas Jefferson, *Letter to James Madison*, Dec. 20, 1787.
3. Albany Inst. of Hist. & Art, *Three Centuries of American Landscape Painting*, *available at* http://www.tfaoi.com/newsmu/nmus92b.htm.

4. *American Eden: Landscape Paintings of the Hudson River School from the Collection of the Wadsworth Atheneum Museum of Art: June 6-Aug. 29, 2004, available at* http://www.tfaoi.com/aa/4aa/4aa360.htm (quoting Lawrence J. Wheeler, director, North Carolina Museum of Art).
5. Yellowstone Park Act of 1872, 17 Stat. 32, 16 U.S.C. § 21 (1872).
6. William Poole, *Conservation and Patriotism: A Conversation with William Cronon, available at* http://www.tpl.org/tier3_cd.cfm?content_item_id=21065&folder_id=3308 (quoting William Cronon, University of Wisconsin).
7. Thomas Jefferson, *First Inaugural Address*, Mar. 4, 1801.
8. Village of Belle Terre v. Boraas, 416 U.S. 1 (1974) (upholding village zoning ordinance against challenge that its narrow definition of "family" effectively excluded students and nontraditional living groups from the entire village).
9. Martin Luther King Jr., "I Have a Dream," delivered at the Lincoln Memorial, Washington, D.C., Aug. 28, 1963.
10. *See* 33 U.S.C. § 1251(a)(2) ("it is the national goal that wherever attainable, an interim goal of water quality which provides for the protection and propagation of fish, shellfish, and wildlife and provides for recreation in and on the water be achieved by July 1, 1983").
11. *See* 33 U.S.C. § 1251(a)(1) ("it is the national goal that the discharge of pollutants into the navigable waters be eliminated by 1985").
12. 42 U.S.C. § 4331(b)(1).
13. 42 U.S.C. § 4331(b)(4).
14. *See* Julie Sturgeon & Janice Meer, *The First Fifty Years (1956–2006): The President's Council on Physical Fitness and Sports Revisits Its Roots and Charts Its Future.*
15. To hear the "Chicken Fat Song" (with its enthusiastic refrain, "Go you chicken fat, go away! Go you chicken fat, go!"), *see* http://www.jfklibrary.org/Asset+Tree/Asset+Viewers/Audio+Video+Asset+Viewer.htm?guid=–C7023694-DB32–49C6-A8F2–5039B8626E40}&type=Audio.
16. Although Wonder bread became America's first sliced bread in 1925, the Strong Bodies campaign flourished in the 1960s. *See About Wonder*, *available at* http://www.wonderbread.com/history.html. For representative ads, *see Helps Build Strong Bodies 12 Ways!* Flickr, *available at* http://www.flickr.com/photos/jikan/3358785303/ (1967 advertisement); *Wonder Helps Build Strong Bodies*, Flickr, *available at* http://www.flickr.com/photos/felixtcat/2523541315/ (1965 advertisement).
17. Sturgeon & Meer, *supra* note 14, at 46.
18. 42 U.S.C. § 7409(b)(1).
19. 33 U.S.C. § 1312(a).
20. 42 U.S.C. § 4331(b)(2).
21. Jonathan Mason, *Patriotism a Virtue* (Mar. 5, 1780), *in* AMERICAN PATRIOTISM: SPEECHES, LETTERS AND OTHER PAPERS WHICH ILLUSTRATE THE FOUNDATION, THE

DEVELOPMENT, THE PRESERVATION OF THE UNITED STATES OF AMERICA 125–34 (Selim H. Peabody, ed., Am. Book Exchange 1880).

22. Nat'l Archives, *Powers of Persuasion: Poster Art from World War II, available at* http://www.archives.gov/exhibits (follow links to "powers of persuasion" and to "He knew the meaning of sacrifice!").
23. *Id.* (follow links to "power of persuasion" and "use it up, wear it out").
24. 16 U.S.C. §§ 529, 531(b).
25. 16 U.S.C. § 1531(b).
26. *See* http://it.stlawu.edu/~quack/seminar/home.htm.
27. President Reagan, First Inaugural Address (1981), *available at* http://www.yale.edu/lawweb/Avalon/presiden/inaug/reagan1.htm.
28. *See* 42 U.S.C. §§ 7651–7651(o) (1990).
29. EPA, *The Plain English Guide to the Clean Air Act: Reducing Acid Rain, available at* http://www.epa.gov/air/caa/peg/acidrain.html.
30. THE ECONOMIST, July 6, 2002.
31. Envtl. Def. Fund, *The Cap and Trade Success Story*, *available at* http://www.environmentaldefense.org/page.cfm?tagID=1085.
32. *See, e.g.*, Lesley K. McAllister, *Beyond Playing "Banker": The Role of the Regulatory Agency in Emissions Trading*, 59 ADMIN. L. REV. 269, 313 (2007) (identifying the limitations of cap-and-trade programs and concluding that "[d]esigners of cap and trade programs that minimize the agency's role and promise a reduction in administrative costs risk sacrificing environmental outcomes to an idealized version of how cap and trade regulation works"); David M. Driesen, *Trading and Its Limits,* 14 PENN ST. ENVT'L L. REV. 169, 172 (2006) (identifying prerequisites for successful environmental trading programs and concluding that, "[w]hen these requisites are not present, trading often proves very problematic").
33. Mikhail Gorbachev resigned as president of the Soviet Union on December 25, 1991, and the Supreme Soviet presided over its own dissolution and acknowledged the termination of the union on December 26, 1991.
34. *See* Depository Institutions Deregulation and Monetary Control Act, Pub. L. No. 96–221 (deregulating the industry).
35. *See, e.g.*, Christopher Dodd, *Hearing on Mortgage Market Turmoil: Causes and Consequences*, U.S. S. Comm. on Banking, Housing, and Urban Affairs, Mar. 22, 2007.
36. Exec. Order No. 12,291 (Feb. 17, 1981), at § 2(b)–(d).
37. *See, e.g.*, Emergency Planning and Community Right-to-Know Act of 1986, 42 U.S.C. §§ 11,001–11050.
38. Pub. L. No. 106–554.
39. *See* OMB, *Validating Regulatory Analysis: 2005 Report to Congress on the Costs and Benefits of Federal Regulations and Unfunded Mandates on State, Local and Tribal Entities, available at* http://www.whitehouse.gov/omb/inforeg/2005_cb/final_2005_cb_report.pdf (estimating annual benefits at $44.3 billion to $233.7 billion, and estimating annual costs at $15.1 to $16.7 billion). The report found that the

"majority of the large estimated benefit of EPA rules is attributable to reduction in public exposure to a single air pollutant: fine particulate matter." *Id.*

40. David Brooks, *The Happiness Gap* (Op-Ed), N.Y. TIMES, Oct. 30, 2007.
41. *See* U.S. Climate Action Partnership, *available at* http://www.us-cap.org/.
42. *See generally* Robert C. Illig, *Essay: Al Gore, Oprah, and Silicon Valley: Bringing Main Street and Corporate America into the Environmental Movement,* 23 J. ENVTL. L. & LIT. 223 (2008).
43. President G.W. Bush, *Remarks at the 2002 President's Environmental Youth Awards Ceremony*, *available at* http://whitehouse.gov/kids/connection/20020418–8.html.
44. Senator Barack Obama, Speech at the Democratic National Convention, Aug. 28, 2008 (accepting the Democratic presidential nomination) (noting that "government must lead on energy independence, but each of us must do our part to make our homes and businesses more efficient"); President Barack Obama, Inaugural Address, Jan. 20, 2009 (pledging to "harness the sun and the winds and the soil to fuel our cars and run our factories").
45. Thomas Friedman, *The Power of Green,* N.Y. TIMES MAGAZINE, Apr. 15, 2007. *See also* RICHARD NELSON ET AL., PATRIOTISM AND THE AMERICAN LAND (Orion Soc'y 2002).
46. Friedman, *supra* note 45.
47. *See* Civil Soc'y Inst., *Who We Are*, *available at* http://www.civilsociety institute.org/who.cfm (click on "CSI Campaigns, Red, White and Green!"). The Society describes itself as a catalyst for change in the areas of regenerative medicine, climate change, and economic change, seeking out "examples of creative thinking and activities already underway." *Id.*
48. http://www.40mpg.org/pdfs/031705_40mpg_survey_report.pdf. *See also* Shawn Langlois, *Shifting Gears: New Kind of Patriotism Drives Americans to Efficient Autos,* WALL ST. J., MarketWatch, Mar. 18, 2005.
49. Nobel Peace Prize, *Press Release*, Oct. 12, 2007, *available at* http://nobelprize.org/nobel_prizes/peace/laureates/2007/press.html.
50. Kent Sullivan, *$25 Million Offered in Climate Challenge: Tycoon Hopes to Spur Milestone Research*, WASH. POST, Feb. 10, 2007 at A13.
51. *See* http://www.iccr.org/.
52. *See* http://www.iccr.org/issues/globalwarm/.
53. *See* http://www.whatwouldjesusdrive.org/intro.php (explaining that WWJDrive "is organized and sponsored by the Evangelical Environmental Network (EEN), a biblically orthodox Christian environmental organization").

Index